Climate Change and Water Scarcity in the Middle East

T0271170

As water's significance as a geopolitical resource is poised to surpass that of oil, this book explores the adaptation of Water, Sanitation, and Hygiene (WASH) services in the Middle East to climate change challenges, leveraging the Humanitarian-Development-Peace nexus for a sustainable transition and resilient solutions. Delving into the humanitarian and development sectors across the region, the authors advocate for a transformative approach towards more innovative, integrated, and localized programming. It draws a parallel between the increasing global shift in humanitarian needs, as starkly revealed by the COVID-19 pandemic, and the ongoing devastation wrought by climate change, particularly through water-related crises such as flooding, drought, famine, and conflict.

The authors stress the urgent need for adaptive and sustainable strategies that can swiftly respond to evolving climate challenges. This book argues that there is currently a window of opportunity for WASH practitioners to develop broader, multi-sectoral experiences to meet these challenges. Drawing on discussions with humanitarian and development practitioners and new contemporary case studies, this book analyzes the financial, institutional, environmental, technical, and socio-cultural considerations for creating sustainable WASH services in transition. The narrative emphasizes the urgent need for a Humanitarian-Development-Peace nexus approach, advocating for multisectoral collaboration and localization as vital to addressing protracted crises and climate change's escalating threats. It calls for a strategic shift towards organizations that merge immediate humanitarian aid with sustainable development, enhancing local capacities for effective, enduring solutions. The authors conclude by outlining practical actions for humanitarian and development organizations at the local, national, regional, and global levels to support effective integrated and transitional WASH programming in the future.

Mariëlle Snel is a senior Global Humanitarian WASH advisor for Save the Children International at the regional Middle East/Eastern Europe office based in Amman, Jordan. She has over 25 years of WASH experience in designing, implementing, researching, and publishing around contemporary issues including the humanitarian-development-peace WASH nexus, conflict, COVID, and climate change over the humanitarian, transitional, and development divide.

Nikolas Sorensen is a humanitarian and development practitioner and researcher who has consulted with organizations like Save the Children US and Plan International. He has previously published works on the humanitarian-development-peace WASH nexus, specifically looking at the challenges of climate change, COVID-19, and conflict.

Reed Power is a humanitarian and development practitioner, most recently working as a senior program officer at World Vision International, based in Iraq. His experience has included a strong focus on WASH linked to climate change and the humanitarian-development-peace WASH nexus.

Earthscan Studies in Water Resource Management

The Role of Law in Transboundary River Basin Disputes
Cooperation and Peaceful Settlement
Chukwuebuka Edum

Desalination and Water Security
Chris Anastasi

Flood Risk and Community Resilience
An Interdisciplinary Approach
Lindsey Jo McEwen

Water Politics and the On-Paper Hydropower Boom
Power, Corruption, and Sustainability in Emerging Economies
Özge Can Dogmus

Water Justice and Groundwater Subsidies in India
Equitable and Sustainable Access and Regulation
Gayathri D. Naik

Climate Change and Water Scarcity in the Middle East
A Transitional Approach
Mariëlle Snel, Nikolas Sorensen and Reed Power

For more information about this series, please visit: www.routledge
.com/books/series/ECWRM/

Climate Change and Water Scarcity in the Middle East
A Transitional Approach

Mariëlle Snel, Nikolas Sorensen and Reed Power

Routledge
Taylor & Francis Group

LONDON AND NEW YORK

First published 2025
by Routledge
4 Park Square, Milton Park, Abingdon, Oxon OX14 4RN

and by Routledge
605 Third Avenue, New York, NY 10158

Routledge is an imprint of the Taylor & Francis Group, an informa business

British Library Cataloguing-in-Publication Data
A catalogue record for this book is available from the British Library

ISBN: 978-1-032-56670-2 (hbk)
ISBN: 978-1-032-56673-3 (pbk)
ISBN: 978-1-003-43670-6 (ebk)

DOI: 10.4324/9781003436706

Typeset in Times New Roman
by codeMantra

Contents

Abbreviations

ACF	Action Against Hunger
ACWUA	Arab Countries Water Utilities Association
FAO	Food and Agriculture Organization of the United Nations
FIETS	Financial, Institutional, Environmental, Technical, and Social
GWC	Global WASH Cluster
GWP	Global Water Partnership
GWR	Groundwater Relief
HCF	Healthcare Facilities
HDP	Humanitarian, Development and Peace
HRP	Humanitarian Response Plan
ICRC	International Committee of the Red Cross
IDP	Internally Displaced Person
IHL	International Humanitarian Law
IOM	International Organization for Migration
JOF	Joint Operational Framework
LRRD	linking relief, rehabilitation, and development
MBR	Membrane Bioreactor
MENA	Middle East and North Africa
NRC	Norwegian Refugee Council
NWOW	New Way of Working
OECD	Organization for Economic Co-operation and Development
SDGs	Sustainable Development Goals
SWA	Sanitation and Water for All
UNHCR	United Nations High Commissioner for Refugees
UNICEF	United Nations International Children's Emergency Fund
WASH	Water, Sanitation, and Hygiene
WEF	Water, Energy, and Food
WHO	World Health Organization
WSS	Water Supply and Sanitation

1 Climate Change in the Middle East

Climate change is the great looming threat of the next century. The challenge of climate change impacts every part of life, and it "affects everything from geopolitics to economies to migration. It shapes cities and life expectancies" (The Economist 2023). Countries will require significant resources and innovation to navigate this slow-onset disaster's current and future pressures successfully. The impacts of climate change are not only misunderstood, but they are also transcending borders, crossing sectors, and influencing every part of human life. Vulnerable and marginalized communities are set to bear the brunt of climate change's impacts, with the least developed countries facing the greatest uncertainties. Climate change significantly threatens current humanitarian and development practices and norms.

The latest figures suggest that approximately 300 million people require humanitarian assistance, and these numbers are anticipated to grow significantly over the next decade if recent trends continue (OCHA 2023b). Conflict, violence, and instability are on the rise, and in 2022 alone, nearly 110 million people were displaced from their homes as a result (UNHCR 2023). The longevity of these crises adds another layer of complexity. For instance, the average humanitarian appeal in 2018 spanned nine years, nearly doubling from the five-year average in 2014. This pattern of protracted crises can be seen across the Middle East and Africa in countries like Syria, Somalia, Yemen, the Democratic Republic of the Congo, South Sudan, and Sudan and currently consume the majority of humanitarian resources globally (Development Initiatives 2023). By 2030, it's projected that over 80% of the world's poorest will live in such fragile contexts, posing immense challenges to traditional humanitarian responses, including the vital field of water, sanitation, and hygiene (WASH), which is critical to addressing climate change in the Middle East (UNICEF 2019b).

Now, two decades into the 21st century, we are seeing rising global conflict and instability, leading to increased forced migration. One of our generations' most significant challenges lie in addressing the growing complexity of an expanding, mobile population and the rising conflict associated with dwindling natural resources. These challenges are only amplified by the impact of

DOI: 10.4324/9781003436706-1

global climate change, which particularly impacts marginalized populations living in areas of poverty and instability. This is particularly true in the Middle East and North Africa (MENA), where the effects of climate change and water scarcity are poised to become monumental challenges for both local and national governments in the coming century. The international community, most notably the United Nations, has emphasized the devastating impact climate change will have on depleted regional water supplies and food production, in addition to the potential catalyst for terrorism, violent extremism, and cross-boundary conflict (United Nations 2020). Over the next half-century, the entire region will enter a period of exacerbated fragility, from the wealthier Gulf nations to those in conflict or undergoing post-conflict transitions. The MENA region is already experiencing the signs of the impact of climate change, with average temperatures rising faster than the global rate (Alaaldin 2022). This shift aggravates existing challenges of weakened governance, demographic pressures, economic instability, and the repercussions of recent pandemics. The Intergovernmental Panel on Climate Change predicts a temperature increase of 1.5°C–2°C in the MENA region by 2040, exacerbating water scarcity, heatwaves, droughts, and dust storms, thereby threatening food security and amplifying poverty and social unrest (IPCC 2018).

In response, humanitarian and development practitioners increasingly focus on integrating disaster risk reduction and climate change adaptation into their strategies. This approach acknowledges that disasters are more than disruptive events in the development journey that can be resolved solely by swift emergency responses. Instead, they stem from unaddressed risks inherent in the development process. Disasters occur when events like floods or earthquakes impact areas where people, property, and systems are inadequately protected. However, we can significantly mitigate disaster risks by implementing strategies that minimize vulnerability and the risk of exposure to such hazards and by addressing broader issues like poverty and inequality. This concept, known as disaster risk reduction, involves crafting humanitarian efforts that address immediate crises and safeguard fundamental human rights in the immediate and long-term aftermath (Turnbull, Sterrett, and Hilleboe 2013). Turnbull, Sterrett, and Hilleboe articulated this further, stating that,

> to reduce disaster and climate change risk, exposure needs to be minimized, vulnerability reduced, and capacities for resilience strengthened in ways that address both disaster and climate change risk simultaneously, neither approach compromising the other. This is a dynamic process requiring continual effort across economic, social, cultural, environmental, institutional, and political spheres to move from vulnerability to resilience.
> (Turnbull, Sterrett, and Hilleboe 2013, 8)

Successfully addressing climate change requires a multisectoral merging of necessary, short-term mitigation and adaptation programing with long-term resilience building and sustainability frameworks, thus grounding climate change response firmly in humanitarian and development actors' mandates.

Yet, the unfortunate reality is that climate change disproportionately affects disadvantaged communities and, most notably, younger populations under the age of 25, who make up half of the entire population of the MENA region (OECD 2022b). The younger generation is forced to deal with the consequences of this insurmountable burden they had no role in bringing about (Alaaldin 2022). In the 2019 UNICEF report Water Under Fire, it was noted that In protracted crises, the risk of death for children under 15 from diarrheal diseases due to inadequate WASH services is nearly three times that of dying from direct conflict-related violence. This risk is even more pronounced for children under 5, who are over 20 times more likely to succumb to such diseases than to violence. These realities highlight the critical need to promptly and adequately deliver WASH services in these situations and make it vital for the short-term and long-term survival of the most vulnerable. Regrettably, humanitarian actors often face resource shortages that require them to rely on underdeveloped and unsafe water and sanitation infrastructure (UNICEF 2019a). Water scarcity and climate change compound these issues in the Middle East, increasing the need for a stronger, more integrated response.

Climate Change Dynamics and Conflict

The Middle East is a region currently facing various geo-political and environmental challenges, including the increasing pressures of climate-related water scarcity. These pressures are further exacerbated by regional population growth and economic development (Lelieveld et al. 2012). Internal conflicts and the lack of long-term programing based on sustainability and resilience have considerable consequences, both now and in the future. The consequences include decreasing water resources, soil fertility for food production, land management, and effects along coastal areas and marine resources (Cloern et al. 2016). Data from the region indicate that water scarcity coupled with population growth is greatly increasing the demand for water in the region (Lelieveld et al. 2012; Stavi et al. 2022).

Agriculture accounts for approximately 70% of water use globally, which is a particular issue in the Middle East (Penney and Muyskens 2023). Even in areas where yearly rain and snowfall averages remain consistent, climate-related droughts and floods are still significantly impacting agriculture, and climate change adaptation in the industry is necessary to meet future needs. Penney and Muyskens point to a growing trend of inconstant rainfall, often coming all at once, requiring farmers to rely more on water storage and irrigation than on an even spread of rain over a growing season (Penney and Muyskens 2023). Combining these problems with climate change, desertification, deforestation, overall land degradation, and air pollution, it is clear the

Middle East faces a long, uphill battle to mitigate current shocks and build adaptive, sustainable, and resilient services capable of meeting future needs (Bayram and Gökırmaklı 2020).

So, how can local, national, and international humanitarian and development actors effectively deal with these issues? The Middle East's complex history and rich sociocultural, environmental, and educational developments have created significant potential for innovations that promote sustainable innovations to these ecological challenges. Biodiversity-friendly farming schemes that increase habitat heterogeneity and the presence of islands or corridor reserves can support local wildlife and minimize the impacts of agriculture on wildlife species. Habitat utilization patterns are examples of progress (Baker et al. 2017). Furthermore, environmental commitment, which involves minimizing waste, using environmentally friendly products, and adhering to governmental politics, has enhanced individual attitudes toward the environment and promoted environmental friendliness (Cop, Alola, and Alola 2020). These, coupled with the increased focus on effective local, national, and regional resource management and continued discussions around sustainability and resilience building, can help Middle Eastern countries, decrease conflict and resource waste, and alleviate further social, economic, and environmental degradation.

Conflict is a significant driver of crisis. Indeed, more countries are experiencing violent conflict now than at any other time since 1989 (Cantor 2023). In conflict, deliberate and indiscriminate attacks destroy infrastructure, injure personnel, and cut off the power that keeps vital systems running (United Nations 2018). These attacks negatively impact the long-term health and economic livelihoods of individuals, families, and communities living in fragile states. Armed conflict also limits access to essential equipment and consumables such as fuel or chlorine, which can be depleted, rationed, diverted, or blocked from delivery. Very often, essential services are intentionally denied or destroyed, particularly in cities, where communities depend on a complex, interconnected set of services. Such attacks on water, sanitation, and power systems can be instantly debilitating and come with long-term consequences, especially for the poorest and most marginalized groups in societies.

Environmental degradation and climate change impacts significantly compound the issues posed by conflict and have far-reaching implications for human security. Limiting access to the resources necessary for human survival reduces health systems' capacities and increases people's health needs, disrupting livelihoods and reducing the adaptive capacities of individuals, families, and communities. During our interview with Karine Deniel, she stated, "Environmental impacts are many times felt cross-boundary, in that they can exacerbate conflicts, disease outbreaks, and negatively impact livelihoods not only within, but between, countries in the Middle East" (Deniel 2023). In addition, when armed conflict is combined with environmental degradation and climate change, it can lead to a significant increase in displacement, threatening human security. People who are displaced due to armed conflict, especially those in informal settlements, are particularly vulnerable to climate-related shocks and disasters. A changing climate can harm human security by damaging livelihood systems or the infrastructure and ecosystems that support them (ICRC and NRC 2023).

With these impacts in mind, there is a critical lack of climate-related humanitarian adaptation support for displaced populations or populations at risk of displacement in conflict-affected countries (Sitati et al. 2021). Displacement can be prevented by providing environmental management and climate adaptation support to vulnerable communities already bearing the consequences of armed conflict before they exhaust existing adaptation options or are exposed to extreme weather events. These actions strengthen resilience. Similarly, there is a need for adaptation initiatives geared towards displaced people to help prevent further displacement, especially for those living for protracted periods in camps or informal settlements – built initially as short-term solutions – and thus lack vital reliance contingencies, making them particularly vulnerable to climate risks. Additionally, it is essential to recognize that both in-country mobility and cross-border mobility can be a critical coping adaptation or survival strategy for marginalized groups facing the combined effects of climate change, environmental degradation, and armed conflict. As such, mobility-related considerations should be integrated into adaptation support strategies and approaches, considering the priorities and concerns of those affected (ICRC and NRC 2023).

Across the Middle East, conflicts have been growing in number and length as climate-related water scarcity drives increased conflict over increasingly scarce resources in a region facing rapid urbanization and population growth. Ensor (2011) argued that conflict is gaining more attention in debates around climate change as the realities that climate change poses to scarce resources, particularly water, are only projected to worsen (Ensor 2011). The pressures on water scarcity are particularly acute in the Middle East and lead to the increased migration of marginalized groups, who often lack the adaptive capacity to meet the unique challenges caused by climate change. These migrations, often into cities, place additional stress on weak water infrastructure and WASH services that were never designed for the increased population and lack critical resiliency to current and future anticipated pressures. Migration can also stress areas previously considered safe and create feedback loops that generate downward spirals of resource competition and conflict that sustain ongoing and protracted violence (Wehrey et al. 2023). Long-term institutional weakness has a lasting impact on social trust and can lead to further degradation in security, livelihoods, unpredictable and changing needs across socioeconomic groups, and political instability (Mosel and Levine 2014).

The Role of the Humanitarian, Development, and Peace Nexus

The humanitarian, development, and peace (HDP) nexus refers to the integration of short-term humanitarian and long-term development approaches with local and national institutions as a means of helping crisis-affected populations. This transition involves a shift from solely providing short-term emergency relief towards longer-term sustainable development programing that addresses the underlying causes of protracted crises. This transition aims to improve the efficiency and effectiveness of programs by promoting collaboration, reducing duplication, and addressing challenges that block progress along the humanitarian and development continuum.

At the World Humanitarian Summit 2016, the development of the Grand Bargain and the New Way of Working (NWOW) cemented the need for greater integration and cooperation between humanitarian and development actors. The reality is that global humanitarian crises were growing in length and scope, significantly altering previous conceptions around a short, linear transition between humanitarian and development response. As crises become increasingly protracted, a shift in thinking becomes critical to future disaster and crisis response (Center on International Cooperation 2019; IASC 2021; ICVA 2017; Nakamitsu et al. 2017; OCHA 2017; United Nations 2017). Following the adoption of these new conventions, humanitarian and development actors globally began reevaluating existing frameworks and decision-making structures to allow better integration. The Global WASH Cluster (GWC) in 2019 published its WASH Roadmap 2020–2025, highlighting essential steps linking WASH services to greater social and health outcomes (Global WASH Cluster 2019). In fact, there are strong linkages between WASH services and other sectors like health, nutrition, shelter, education, nutrition, and child protection. These connections make water resource management and WASH programing a critical starting point for addressing the larger outcomes each of these individual sectors wishes to address (Snel and Sorensen 2023b). Yet we note here that the impact climate change has on water, water scarcity, and, by default, WASH services compounds multisectoral outcomes, necessitating greater, multisectoral coordination to create effective and sustainable resilience for at-risk populations, particularly in already water-scarce areas like the Middle East.

More specifically, the humanitarian and development nexus transition is a critical concept in the context of the WASH sector in the Middle East. The region has faced numerous crises in recent years, including conflict, displacement, and natural disasters, which have severely impacted WASH services and infrastructure. As such, successfully transitioning from humanitarian to development programing is essential to ensuring sustainable access to clean water, adequate sanitation, and good hygiene practices in the region. However, this transition is not without its challenges. The region faces unique political, social, and economic challenges that can hinder the implementation of sustainable and resilient WASH programing. These challenges include limited funding, weak governance, ongoing conflict and instability, and the impact of climate change (OECD 2022a). Effectively addressing these challenges will be essential to achieving the long-term goal of sustainable access to WASH services and infrastructure for all.

The Strengths-based Approach, the Joint Operational Framework, and the Sustainable WASH Model

Although numerous shared principles broadly guide the efforts of humanitarian and development organizations in the WASH sector, there is a lack of alignment in their approach to planning, designing, and implementing interventions, which results in prolonged and ineffective transitions. Shocks and stresses such as climate change, conflict, and COVID-19 have devastated water resource management and WASH services in vulnerable countries worldwide. This is particularly true in the Middle East, where water scarcity drives many local, national, and regional conflicts (Abel et al. 2019; Baxter et al. 2022). Additionally, peace-related aspects are frequently overlooked. This dissonance in the sector coincides with a rise in the frequency and duration of humanitarian crises, impacting larger populations, especially in areas susceptible to protracted and recurring crises (Mason and Mosello 2016). United Nations High Commissioner for Refugees (UNHCR) provides the first guidance manual reflecting on WASH targets in humanitarian, transitional, and protracted crises as a result of growing humanitarian needs and increasing socioeconomic instability (Harvey et al. 2019). Developing effective, integrative, and coordinated responses to climate change, water scarcity, and WASH services that result in sustainable, long-term resiliency for individuals, families, and communities is a work in progress. In the last few years, two critical frameworks have emerged: a strengths-based approach that empowers local decision-makers and communities and the Joint Operational Framework (JOF), which facilitates coordination among HDP actors and institutions. These frameworks prioritize community empowerment, asset mobilization, and strategic coordination, offering pathways to mitigate climate impacts on and enhance community resilience (Grieve 2023a; Grieve, Panzerbieter, and Rück 2023; Winterford, Rhodes, and Dureau 2023).

Strengths-based approaches have been a common concept for decades in many other fields and contexts but have only been applied meaningfully to international development recently (Winterford, Rhodes, and Dureau 2023). Traditional humanitarian and development frameworks look at areas of underdevelopment or crisis through the lens of deficits, allowing need assessments to identify the resources and skills a given person, group, or community is lacking and thereby simplifying response as filling those deficits.

A strengths-based approach flips the narrative and instead focuses on the assets individuals, families, and communities bring to either short-term humanitarian crises or to long-term development programing. It puts local decision-makers in as the drivers of positive change by allowing them to leverage their assets and capabilities to solve complex problems. We must clarify here that deficits are a reality in humanitarian and development contexts, particularly with marginalized groups who often do, in fact, lack the resources, skills, and capabilities to adapt to the unique stresses during disaster effectively. A strength-based, asset-focused approach, however, places individuals, families, and communities, particularly marginalized groups most impacted during a crisis, as the dividers of deciding what assets they have, what deficits need, and how local, national, and international humanitarian and development resources can be utilized to build sustainability and resiliency for the future.

What makes a strengths-based approach unique is how well it connects with other foundational humanitarian and development mandates and philosophies. Winterford, Rhodes, and Dureau (2023) highlight notable clear connections and synergies between a strengths-based approach and other frameworks like Amartya Sen's capabilities approach with its focus on individual freedoms (Sen 1985, 2000, 2001), The Grand Bargin's renewed push for greater localization (IASC 2021; World Humanitarian Summit 2016) and participatory and sustainable livelihoods approaches as developed by Robert Chambers and others (Chambers 1997, 2006, 2017; Chambers and Conway 1992). Yet it is in light of the growing complexity found in increasing numbers of protracted crises that give a strengths-based, asset-focused approach its potential. Winterford, Rhodes, and Dureau stated that

> A strengths-based approach does not deny inequalities, injustices, and problems: it offers an alternative perspective on how these issues can be addressed. It seeks to address these through an orientation and focus on action towards preferred futures, rather than defining needs, problems-solving, and filling gaps.
>
> (Winterford, Rhodes, and Dureau 2023, 19)

Or, as we argue here, as complexity increases, outcomes become increasingly difficult to achieve and require a shift in focus away from deficits and towards assets, away from ends and towards means.

A strengths-based approach focuses on the current means of achieving desired outcomes by empowering individuals, families, and communities to become leaders and decision-makers in creating their own sustainable and resilient futures. This process breaks traditional and frankly simpler humanitarian and development action models that rely on a linear progression or transitions from disaster and crisis response to stability and flourishing. These traditional models are simpler for humanitarian and development actors to plan and implement as they rely on a top-down orientation for funding, program design, and implementation and are, therefore, easier to achieve in short-term funding cycles. Though potentially slower, the strength-based, asset-focus approach builds a stronger long-term foundation for resilience building and sustainability in program outcomes but comes with a critical loss in control by national, regional, and international policymakers. Finding a balance between the top-down resources and bottom-up program design and control is critical to addressing the growing complexities of climate-related water scarcity in the Middle East.

The JOF, which the German WASH Network commissioned with oversight from UNICEF, Sanitation and Water for All (SWA), and the GWC, was released in early 2023 (Grieve 2023a; Grieve, Panzerbieter, and Rück 2023). Though primarily focused on the WASH sector, the JOF is a powerful tool to help HDP actors increase effective coordination across their separate mandates while also ensuring increased reliance on the input of local stakeholders in the development and implementation of WASH services. The JOF assists policymakers, coordinators, and practitioners at the national and sub-national levels to integrate resilience, conflict sensitivity, and peacebuilding capabilities into existing and new WASH programs by leveraging the nexus approach. This integration enables WASH programs to achieve sustainable development, address and reduce humanitarian WASH needs, and contribute to building resilient, inclusive, and peaceful societies. Applying the JOF is particularly relevant in contexts of protracted and recurrent crises.

In this book, we utilized the Sustainable WASH Model, also known as the Financial, Institutional, Environmental, Technical, and Social (FIETS) model, as an analytical lens (WASH Alliance International 2021). Based on lessons learned in the Middle East, the Sustainable WASH Model integrates five key sub-sectors to ensure the sustainability and effectiveness of WASH initiatives. Financial considerations focus on self-sufficiency, underpinned by local funding mechanisms such as taxes and fees versus the need for external subsidies. Institutional considerations aim for clearly defined roles and cooperative engagement among local and national stakeholders, aligning with user needs and promoting transparent governance. Environmental consideration advocates for the harmonious management of natural water and waste systems, recognizing the interdependence of human activities and ecological health. Technical considerations seek to ensure the continual upkeep and modernization of WASH infrastructure through local stewardship. Lastly, social considerations aim for community-led, equitable, and culturally aware practices supporting robust and healthy societies. Using these frameworks and models, we aim to promote stronger alignment within the WASH sector and the larger challenges of climate change and water scarcity. This focus requires greater consistency between the HDP sectors. It also underscores the importance of upholding organizational responsibilities and recognizing that solutions must be tailored to the specific context and driven by local considerations. The strengths-based approach, the JOF, and this book aim to enhance climate change interventions' efficiency, effectiveness, sustainability, and resilience in humanitarian and development contexts.

Methodology and Findings

In preparing for the book, we utilized the Sustainable WASH Model, as discussed above, as part of a series of qualitative surveys and interviews with WASH practitioners and experts across both the humanitarian and development fields working within a wide range of roles in the WASH sector, including with local, national, regional, and global public and private organizations. Questions were aimed at identifying barriers to effective water resource management and WASH services and, critically, at identifying key innovations and developments to effective climate response. We also sought insights into transitions across the humanitarian-development nexus and the changes needed to facilitate greater integration in regional responses to climate change, water scarcity, conflict, and crisis response. In total, we received 17 qualitative survey responses and conducted 14 interviews. All written and verbal feedback from respondents and interviewees during the survey and interview stage was later transcribed and coded into sub-themes under each core challenge of the core challenges outlined within the Sustainable WASH Model. The following chapters discuss key surveys and interview findings at greater length. We draw on insights from the strengths-based approach and the JOF as lenses to understand findings better and suggest practical future guidance for effective climate change mitigation and adaptation initiatives across the region.

We acknowledge the limitations and bias in this publication. This publication is focused primarily on climate change and water scarcity in the Middle Eastern context and could overlook key environmental debates over climate change, such as pollution. Instead, we have focused our discussion on the challenges that climate change will pose for regional actors and stakeholders. We highlight solutions decision-makers can take to better mitigate and adapt to current and future climate-related pressures. This publication also aims at a specific audience, notably humanitarian and development practitioners who lead the HDP nexus transition and academic thinkers interested in the HDP nexus. We note that WASH practitioners working in the Middle East are a very small group of highly specialized individuals. Not all of them answered our requests to take part in our research. We sought to interview individuals working on all sides of the HDP nexus and from a broad range of individuals working in local, national, and regional positions. The specific nature of the roles targeted could lead to findings that do not fully capture the broader range of challenges faced in all contexts or by all stakeholders. As such, all suggestions here must be fully contextualized by local thinkers and decision-makers and may not be applicable in all contexts.

Furthermore, as a qualitative study, the ideas expressed represent the opinions of individuals and not larger norms. Though experts in their field, the individuals interviewed are also products of their circumstances and contexts. Their current organizational affiliations and roles could heavily influence them, skewing their response to organizational priorities. As this was not a study of the perspectives of local stakeholders or critically impacted marginalized groups, our findings may not fully capture local contexts and their assets or needs. Finally, we note that these findings represent our interviewees and our opinions and thoughts momentarily. Challenges and contexts change rapidly, particularly in humanitarian settings, and as such, findings can, in some cases, become rapidly outdated.

Using the Sustainable WASH Model as a guide, our survey respondents and interviewees suggested that the greatest challenges to addressing WASH services in the Middle East, ranked from most to least pressing, were institutional, financial, and environmental. Technical and social considerations were deemed important but were not touched on heavily. Interviews were used to probe deeper into the specific institutional, financial, and environmental challenges faced in the WASH sector and how the increasing impacts of water scarcity and climate change in the Middle East complicate WASH programming. Interviewees were also asked to discuss case studies from the region that might highlight potential success stories in mitigating or overcoming these challenges and what role integration across the HDP nexus might play in future adaptation and mitigation efforts. Below, we provide a brief overview of our survey and interview findings. Throughout this book, we draw on the voices of our survey and interview respondents to highlight challenges that connect these voices to ongoing research and to proactively attempt to provide clarity in an integrated path to successful climate change adaptation and resilience building for the future. Below is a summary of the feedback for each challenge based on participant responses and organized according to the FIETs model with no implied ranking of significance.

Financial – Core financial challenges for HDP practitioners include uncertainty and reliance on short-term, external funding that compromises sustainability efforts. Concerns encompass dependency on state subsidies, fluctuating donor and government funding, and corruption, which hinder the financial sustainability needed for climate change resilience. Geo-political instability further complicates these challenges, affecting effective climate-related responses to water scarcity and WASH service needs.

Institutional – HDP actors encounter institutional challenges in the Middle East due to weak governance, fragmented coordination, and the need for climate-specific policy and regulatory reforms. These barriers obstruct the planning and implementation of resilient services. Specifically, unclear institutional mandates necessitate policy changes and regulatory reforms to enhance institutions' capacity to address water scarcity effectively.

Environmental – The lack of public awareness, poor water management, and insufficient cooperation exacerbate cross-border environmental impacts, limiting progress across sectors and diminishing livelihood capabilities. Key environmental challenges in the Middle East include water scarcity, climate change, population growth, and conflict, intensifying resource competition and threatening infrastructure. Environmental degradation and conflict increase human security risks, disrupt livelihoods, and require a greater emphasis on adaptation, resilience, and sustainability efforts.

Technical – Core technical challenges for HDP practitioners in the Middle East include a lack of specialized skills and capacity that hinder project implementation and management. These gaps limit innovation, sustainability of water systems, and adaptation to climate change. The emphasis on community-led programing and localization highlights the necessity to integrate local mechanisms and capacities. The changing climate change landscape calls for a more integrated, resilient, and sustainable approach, demanding advanced skills, nexus-thinking, and cross-sector collaboration.

Social – Social challenges significantly affect climate change programing and the effectiveness of transitions across the HDP nexus. Geopolitical tensions and conflicts, especially over scarce resources like water, disrupt WASH services and increase vulnerabilities, particularly for children and marginalized groups. The region's lack of social cohesion, worsened by long-standing conflicts, undermines trust essential for resilient infrastructure and institutional systems. Insufficient peace-building efforts and conflict-sensitive approaches impede effective, community-led programing. Collaboration across HDP sectors is crucial to address these intertwined social challenges and achieve sustainable water resource management and WASH outcomes.

Conclusion

The most pressing threats, as identified by respondents, span FIETS domains, with institutional, financial, and environmental challenges being particularly prominent. Such insights are crucial in shaping a more nuanced understanding of the obstacles the WASH sector faces due to water scarcity and climate change and how these challenges impact the critical transition between HDP actors in the Middle East. Delving deeper, the interviews offered a platform for experts to ask deep questions, expound on challenges, and provide valuable contexts and examples from the field. In our interview with Farah Al-Basha, she summarized some of the questions this book aims to address, stating,

> It is clear that several countries are having droughts in the region, and there isn't enough being done to plan and prepare for this change. The question for the WASH sector is, where do we start? Is it establishing early warning systems or understanding groundwater levels? Do we start with establishing preparedness plans? What is the level of resistance from the government in addressing climate change? Is it the role of the government or the UN to create preparedness plans? We cannot do much if we do not ask and answer these questions in advance and simply jump to respond when disasters occur. It is too late.
>
> (Al-Basha 2023)

Finding the answers to these questions requires coordination, collaboration, and planning. Although we do not claim to be the ultimate authority on the topic, we aim here to bring together key actors and thinkers capable of pointing future climate action in the right direction. As such, the following chapters draw heavily from the feedback and input of our survey respondents and interviewees, who provide a deeper, hands-on insights into the barriers they face and the innovations and actions they recommend to improve WASH outcomes in the Middle East despite growing pressures of water scarcity and climate change.

In subsequent chapters, we examine challenges identified through practitioner discussions, leveraging published and found case studies to illustrate potential solutions. Chapter 2 delves into the Middle East's specific vulnerabilities to climate change and water scarcity, emphasizing adaptation and mitigation strategies crucial for addressing present and future impacts on WASH services. Chapter 3 focuses on institutional obstacles identified as primary barriers to effective climate response, exploring local, national, and regional reforms to bolster sustainability and resilience efforts. In Chapter 4, we dive deeper into the HDP nexus, highlighting the complexities of climate change, resource scarcity, conflict, and forced migration, necessitating a paradigm shift for effective and sustainable solutions. Lastly, Chapter 5 discusses the need for HDP actors to adopt a new nexus mentality based on a strengths-based approach and the JOF as a better means of designing, implementing, and achieving climate-related WASH outcomes in the Middle East. We also look briefly at ongoing financial debates that perpetuate silos between humanitarian and development sectors, presenting integrative approaches and summarizing key findings to guide practical action. In the appendices, key definitions can be found in a short glossary (Annex 1), a list of our interviewees is provided (Annex 2) along with three new case studies (Annexes 3–5), which are discussed in Chapter 5.

Climate change is poised to significantly worsen ongoing crises in the Middle East, including forced migration, water scarcity, food insecurity, pandemics, and protracted conflicts. To address these issues, policymakers must address these challenges by implementing mitigation, adaptation, and resilience measures that reduce the waste of scarce resources. Critically, a paradigm shift is needed to address water scarcity and the provision of WASH services that can effectively and sustainably transition across the HDP nexus while integrating knowledge and resources from various sectors and actors, locally and regionally. Collaboration and nexus thinking are essential for resilience building in the face of climate change and conflict. Only through a holistic approach can the region overcome these challenges and ensure a future of health, dignity, and prosperity.

2 Impact of Climate Change in the Middle East

Climate change has aptly been described as the most significant challenge we face in the twenty-first century, with significant potential to disrupt economies, exacerbate conflicts, and push the most vulnerable further into the margins (Howard et al. 2016; Malerba 2021). As temperatures continue to increase globally, the frequency of droughts, floods, and forced migration will also increase, straining the capacity of governments and humanitarian institutions. Scholars and aid organizations alike have pointed out that the pressures of climate change will lead to decreases in crop yields, affecting livelihoods and nutrition, as well as increases in water scarcity and conflicts as resources become scarcer (Alaaldin 2022; Baxter et al. 2022; Malerba 2021; Norwegian Red Cross 2019; UNICEF 2020). This is particularly true in the Middle East, which faces a complex and ever-evolving set of challenges intensified by the relentless forces of natural and man-made crises, including climate change. The WASH sector, integral to human health and well-being, is caught in the crosscurrents of humanitarian needs, developmental ambitions, and the imperatives of institutional peacebuilding. As the global community grapples with the profound transformations wrought by climate change, the Middle East stands at the intersection of these challenges, demanding innovative solutions and a fundamental reevaluation of how we address them.

Climate change affects all aspects of life, intensifying environmental depletion, economic strain, and social pressures and challenging development priorities. This is particularly true of water resources and WASH services as climate change will severely impact the supply and demand side of water systems (Batchelor, Smits, and James 2011; UNICEF and GWP 2022). Climate change significantly affects vulnerable groups, including women, children, and people living with disabilities, straining healthcare and education systems and undermining their future well-being and livelihoods. Global water cycles will continue to change, leading to increased flooding and drought. Additionally, as global warming continues to rise, weather and water distribution changes are anticipated to increase in intensity and frequency, leading to greater uncertainty and humanitarian needs (Batchelor, Smits, and James 2011; UNICEF and GWP 2022). A recent report by UNICEF and the Global

DOI: 10.4324/9781003436706-2

WASH Partnership (GWP) stated that "with the world ill-prepared to respond to these risks, this can cause loss and damage, which affects the safe, sustainable and equitable access to water, sanitation and hygiene services" (UNICEF and GWP 2022, 1). Baxter et al. (2022) argue that with an escalation in poverty and lack of food security, the world is grappling with deepening crises in its global supply chain and energy sectors. This situation is compounded by growing social and political unrest, the unfolding consequences of COVID-19, and the increased frequency and severity of climate-related incidents. In this challenging context, humanitarian efforts are strained due to limited funding and an increase in violence targeting aid workers, leading to a more restricted environment for humanitarian activities. In conclusion, Baxter emphasizes, "climate change is a threat multiplier, increasing the risk of climate-related crises, conflict, and displacement" (Baxter et al. 2022, 1561).

The growing impact of climate change particularly impacts poor, marginalized groups in areas characterized by weak institutions and conflict, increasing livelihood and health vulnerability, as well as forced migration (Norwegian Red Cross 2019). In 2019, the Norwegian Red Cross stated, "Natural hazards become more frequent disasters where people are already vulnerable. Where people are marginalized, institutions are weak, and/or conflict has stretched people's coping capacities, the humanitarian consequences of climate change increase" (Norwegian Red Cross 2019, 8). Additionally, Malerba (2021) points out poor and marginalized groups "often lack the economic resources and capital (human, social and physical) that are necessary for adaptive capacity to climate change" (Malerba 2021, 688). The very skills, networks, and resources needed to adapt to the current and coming climate crisis are the ones most critically absent in populations most at risk.

Adequately addressing the complex humanitarian and development problems caused by climate change is no small matter, and climate variability already poses and will continue to pose challenges for both developmental and humanitarian actors. Our interview with Anders Jagerskog stated, "This is a challenge that is here for the long haul for both sectors" (Jagerskog 2023). Looking to the future, in the face of this growing challenge, humanitarian and development actors need to prepare to deal with climate change. Jagerskog argued,

> If the developmental and institutional setup during normal times of stability is not in tune, it will only be exacerbated during times of conflict and climate change. This is already being seen among both host communities, [and] in in-camp settings.
>
> (Jagerskog 2023)

In essence, the unpreparedness of HDP sectors to address the challenges of climate change during times of stability creates a heightened vulnerability already manifesting in the region and impacting the most vulnerable populations. During our interview with Anna Rupert, she noted, "Without plans in place, climate change makes a vulnerable system more vulnerable. There is a growing need for a deeper sense of urgency, which will subsequently add a layer of further complications to our line of work" (Rubert 2023). Urgent action is required to align these sectors with climate resilience strategies to mitigate the increasingly dire consequences of climate change in these fragile contexts.

The Middle East, a region characterized by its arid and water scarce environment, is acutely vulnerable to the adverse effects of climate change, including prolonged droughts, erratic rainfall patterns, and increased water stress (IPCC 2018). These climatic shifts compound existing challenges, altering the roles and responsibilities of actors and affecting the availability and quality of water sources and sanitation infrastructure. Humanitarian crises are becoming more frequent and protracted, while socioeconomic instability continues to rise (OCHA 2023b). The HDP nexus transition has emerged as a critical concept in this context. It underscores the necessity of multisectoral alignment across HDP building efforts to tackle the region's multifaceted problems. This transition aims to harmonize these often disparate approaches while recognizing the local nuances and contextual peculiarities that make each situation unique.

The severity and frequency of severe weather events, including droughts and floods, are rising throughout the Middle East, which contains 12 of the 17 most water-strained nations globally (Alaaldin 2022). A recent example includes the five-year drought in Syria that ultimately led to the Syrian conflict. The floods in Pakistan displaced upward of 33 million people and distilled fear among other nations in the region that they may face similar events in the not-so-distant future (Baghdadi 2022). In Syria, a five-year drought in 2007 led to significant pressures on livelihoods and increased fragility. The subsequent economic impact and influx of refugees caused poverty to skyrocket in the most affected urban areas, leading to a massive wave of urbanization, overburdening already stressed infrastructure and institutions, and leading to civil war in 2011 (Alaaldin 2022). As countries across the Middle East seek to address the impacts of climate-related flooding and drought, the World Bank estimated that the effects of water scarcity on agriculture and healthcare industries alone could cost Middle Eastern countries 6% or more of their GDP by 2050 (World Bank 2016).

If anything is clear regarding climate change, particularly in the Middle East, it is messy, complex, and here to stay. Batchelor, Smits, and James noted that "there is also growing recognition that climate change impacts on both the supply and demand sides of WASH systems" (Batchelor, Smits, and James 2011, 17). Countries across the Middle East will have to anticipate the challenges brought by floods and drought in their efforts to build resilience and sustainably provide access to safe water and sanitation or risk increased health outcomes, environmental degradation, and potential for conflict. Succeeding in such challenges will take innovation and tenacity in both thought and action. Following this line of thought, Caravani et al. (2022) argue for a proactive approach in addressing climate change and crisis settings, stating

> It is time to embrace uncertainty, confront ignorance, and generate reliability through such approaches to professional practices, organizational arrangements, financing, and accountability relations. If not, the failures of mainstream risk-and-control-focused interventions across social assistance, humanitarian relief, and disaster response will persist, and uncertainties in crisis and conflict settings will remain ignored and unaddressed – often with disastrous consequences.
>
> (Caravani et al. 2022, 14)

However, even more than this may be needed. Making this point, Malerba (2021) argues that "Even if bold action is taken and climate change mitigation is adequately addressed, climate change adaptation is still unavoidable" (Malerba 2021, 688). This reality makes action now critical. Addressing climate change requires mitigating the current impacts of rising temperatures, sea levels, water scarcity, and other climate crises while addressing the core sources of these problems by sustainably reducing and mitigating our short-term and long-term environmental impact.

The Sustainable Development Goals (SDGs), developed by the UN in 2015, aim to make substantial developmental progress globally by 2030, including climate change and its impact (United Nations n.d.). There is a strong correlation between SDG 13's mandate to "take urgent action to combat climate change and its impacts" and SDG 6's mandate to "ensure availability and sustainable management of water and sanitation for all" (United Nations n.d.). This leads to a natural connection between climate change reduction and humanitarian and development WASH services. UNICEF's strategy for WASH "indicates that strong engagement is needed to adapt to climate change and follow environmental and social standards, while maintaining the focus on helping every child gain access to WASH, including in schools and health centres, and in humanitarian situations" (UNICEF 2020, 14). Yet a recent UNICEF and GWP report highlighted the intersectionality of climate change resilience services by emphasizing the connection between climate change and SDGs around "poverty, hunger, health, education, affordable and clean energy, gender equality, the reduction of inequalities, sustainable cities and communities, and peace" (UNICEF and GWP 2022, 8). The same report went on to show a particular connection between climate change and the water-food-energy nexus and stated that "the links between the three mean that an integrated approach is required to ensure the security of water and food, and the sustainable production and use of energy worldwide, while also securing ecosystem needs." (UNICEF and GWP 2022, 8). As demand for water, food, and energy continues to increase, the need for stronger integration between climate resilience strategies and the water, food, and energy sectors will be essential to producing sufficient and sustainable resources to meet current and future needs.

There is, however, a recognizable disconnect between the need to address climate change at its root causes and an acceptance of who is ultimately responsible for leadership. A decade ago, Batchelor, Smits, and James pointed out that

> there seems to have been a gradual shift from WASH professionals and practitioners doubting that climate change poses a risk to WASH services delivery, to taking the stance that climate change is a potential hazard, but is "somebody else's problem."
>
> (Batchelor, Smits, and James 2011, 9)

These authors noted a trend for politicians and WASH professionals to blame problems in WASH services on climate change yet simultaneously excluding themselves from climate change research and adaptation plans (Batchelor, Smits, and James 2011). They also reveal a split whereby WASH professionals attribute service failures to climate change but persist with traditional program designs and 'business as usual,' merely nodding to adaptation and resilience strategies without significant changes. This is particularly true in humanitarian settings, whereby WASH actors often delegate climate change responses to long-term development efforts. Meanwhile, traditional WASH services, especially financing, are increasingly integrated with other humanitarian and development sectors by public health, climate change resilience, and sustainability teams. Addressing the complex needs of climate change and WASH services effectively requires enhanced integration across humanitarian and development sectors. WASH, intersecting with healthcare, education, child protection, and more, benefits from this cross-sectoral collaboration, suggesting a future direction for effectively tackling challenges like climate change, water scarcity, and food insecurity.

Regardless of which side of this debate HDP professionals take, immediate action is crucial. We recognize, echoing Batchelor, Smits, and James (2011) – a perspective still relevant today – that in certain regions, immediate threats like dwindling unpolluted water sources take precedence over climate change challenges in the short term (Batchelor, Smits, and James 2011). As such, the all-program design must be contextualized to local priorities. Yet, in the long term, the consequences of climate change will only worsen, and action everywhere is needed now. The unique challenges come from addressing climate change's sporadic and inconsistent nature as it presents itself in sudden onset natural disasters (floods, storms, droughts) and slow-onset disasters (temperature increases and sea-level rises). Attributing causation can be difficult, especially in fragile states prone to conflict. However, in 2019, Abel et al. pointed at the Syrian conflict as a prime example of how climate-related pressures can lead to political unrest, conflict, and civil war. Abel et al. noted that the Syrian conflict grew out of "water scarcity and frequent droughts, coupled with poor water management, [which] led to multiyear crop failures, economic deterioration and consequently mass migration of rural families to urban areas" (Abel et al. 2019, 239). Urban areas in Syria suffered from the continual increase in population, which led to overcrowding, unemployment, and social pressures, which eventually boiled over into political unrest and armed conflict (Abel et al. 2019). The International Committee of the Red Cross (ICRC), however, highlighted a growing new reality that "climate change will make the direct and indirect humanitarian consequences of armed conflict even worse. And, furthermore: That climate change will negatively impact possibilities to end conflicts" (Norwegian Red Cross 2019, 5–6). Climate change can be both the cause of conflict and a significant driving factor preventing peace negotiations.

Current climate change response thinking heavily draws on resilience and sustainability frameworks in humanitarian and development contexts. In 2022, UNICEF and the GWP updated the Strategic Framework for WASH Climate Resilient Development, a framework aimed at "provid[ing] safe, sustainable and climate resilient WASH service delivery to exposed and vulnerable populations, both now and in the future" (UNICEF and GWP 2022, 6). Achieving this objective requires the unified efforts of HDP institutions and organizations to meet current and future needs. There is increasing recognition of the WASH sector's potential to help mitigate the effects of climate change by reducing water and energy waste and emissions while increasing efficiency in resource use and operations. Still, a critical increase in investment is needed to bridge climate-reliance gaps in the sector (UNICEF 2020). UNICEF emphasizes the critical need for enhanced WASH services and additional funding to adapt to climate change, stating

> There is also a specific and pressing need to protect the sector itself, so it can meet the needs of growing populations around the world if the climate changes as predicted in the years ahead. Adaptive capacity will need to be increased at all levels in the coming years; services must be able to continue to function as needed under increased uncertainty and pressures, changing hydrological/hydrogeological conditions, and more frequent extreme weather events. Focusing on these critical elements will help WASH services plan for the future, enhancing their resilience to climate change – and this must be at the core of our WASH programming.
>
> (UNICEF 2020, 5)

A focus on resilience building and sustainability is the bridge needed to prepare for increasing population strain on scarce water resources, especially as the disparities between water supply and demand widen.

Resilience building is closely linked to discussions on integration, referring to the capacity of individuals, families, communities, institutions, and ecosystems to recover efficiently from shocks or disasters. Additionally, resilience dialogue has provided a shared vocabulary and mandate capable of bridging various sectors on both sides of the humanitarian and development divide. In 2015, Tanner et al. noted that "resilience is increasingly providing an integrative 'boundary concept' that brings together those interested in tackling a range of shocks and stresses, including food security, social protection, conflict, and disasters" (Tanner et al. 2015, 23). Resilience is therefore becoming a shared mandate that, when done successfully, has the potential to reduce humanitarian needs while increasing individuals, families, communities, and institution's ability not only to cope with shocks but to thrive after.

Climate change resilience requires a multisectoral approach to address adequately. The Norwegian Red Cross highlighted the direct connection between individual, family, and community coping capacities with the strength of local and national institutions. Areas with weak or weakening institutions lead to decreased livelihood and health outcomes and increased marginalization, conflict, and forced migration (Norwegian Red Cross 2019). Building sustainable climate change resilience requires early institutional strengthening now. It creates the best opportunity for marginalized groups to adapt to coming climate-related shocks while crucially reducing the length and need for humanitarian assistance.

Yet, a recent joint report from the World Bank, ICRC, and UNICEF (2021) argued that despite ongoing discussions about bridging the humanitarian-development divide, not enough is being done between humanitarian and development actors to work together effectively. This is particularly true within the WASH sector and throughout the fragile states and protracted crises in the MENA region. With humanitarian crises growing in length and number, there is a greater need for risk-averse development actors to increase coordination and support of their humanitarian counterparts in order to address the complex WASH needs of affected populations in protracted crises (World Bank, ICRC, and UNICEF 2021). The report further suggests durable solutions are strengthened when development actors proactively initiate humanitarian efforts, particularly in water and sanitation services, during the early stages of a crisis. Likewise, humanitarian teams should prompt development actors to resume projects as early as possible to ensure sustainable recovery efforts continue as they begin to withdraw (World Bank, ICRC, and UNICEF 2021, 59). Granted, in areas of armed conflict, this transition poses greater challenges. However, humanitarian and development actors must continue coordinating to ensure the most effective transition possible while providing a seamless continuation of WASH services.

Water Scarcity and the Environment

The Middle East remains one of the most water-strained regions in the world. Water scarcity will be one of, if not the most pressing, challenges the region faces as it seeks to address the impacts of climate change. Current expectations anticipate the population in the Middle East to double in the next 50–60 years, which will place further pressure on Middle Eastern states to manage resources better, especially water and energy, and seek new, innovative solutions to meet future needs (Waha et al. 2017). Despite mounting future pressures, there is a severe lack of preparation across the region to meet not only current stress but also future challenges. In our recent interview with Jacob Waslander, he stated, "Whether rich countries or poor in the Middle East, the impact of water stress in the upcoming years will be very severe" (Waslander 2023).

It is important to note that the scientific consensus around climate change and water scarcity is not all bad. Howard et al. (2016) point to growing evidence that groundwater will be less impacted by climate change, and groundwater recharge may even increase in some scenarios, making groundwater a key resource in adaptation program design (Howard et al. 2016). One of the major impacts of climate change will be the alteration of precipitation patterns, making rainfall less predictable. This further emphasizes that water resource management will become even more vital as water demand increases with population growth. Successful adaptation in the Middle East will require prioritization of water use and close monitoring of groundwater recharge rates if the region is to meet growing needs. As Waslander stated, "All countries are already having problems with water stress, and this will increase over time. The question is, where do countries and actors start in addressing this?" (Waslander 2023).

Countries in the Middle East have coped with water scarcity for years, says Waha et al., through "abstraction of groundwater, water harvesting and storage, wastewater reuse, desalinization plants, and food imports. The region is at present already highly dependent on food imports" (Waha et al. 2017, 1624). Food importation has been essential as decreases in internal agriculture production significantly reduce water drain but make these countries heavily reliant on international food supplies, reducing self-sufficiency and making them susceptible to shocks in the global supply chain (Waha et al. 2017). Despite these initiatives, a recent report by the World Bank, ICRC, and UNICEF highlights that poor resource management has led to issues not only with water quantity but also with water quality "disproportionately felt among lower-income countries, which cannot afford energy-intensive, high-cost technologies such as desalination" (World Bank, ICRC, and UNICEF 2021, 10). An interviewee, Esmail Ibrahim, observed a general lack of control over water resources in the region, especially during droughts and floods (Ibrahim 2023). He went on to state that "both humanitarian and development actors, as well as governments, cannot continue to construct more and more pumps as a solution. Systems are also outdated and cannot be upgraded to face climate challenges" (Ibrahim 2023). The region is ultimately unprepared in its plans, measurements, catchment, and systems to address climate change's current and future impacts.

Serious management changes are needed to mitigate current waste and address future needs. Even as water scarcity is hitting an all-time high in the region, the wealthy gulf states, Waha et al. (2017) point out, "consume more water per capita than the global average, with Arab residential water and energy markets among the most heavily subsidized in the world. Electricity consumption per capita is twice as high as or higher than the world average," which is a strong indicator that the states with the greatest means to meet the challenge of climate change are underprepared to do so (Waha et al. 2017, 1624). Yet water scarcity will have a greater impact on marginalized groups and countries with the least capacity to adapt to shrinking resources. Decreased water access is also directly linked to more waterborne diseases and lower school attendance, adversely affecting women and girls' education and future livelihoods, as they often fetch water (Corwith and Sorensen 2023; UNICEF and GWP 2022). The consequences of poor management of water resources across the Middle East will tremendously impact marginalized groups in the region.

Addressing poor water management in agriculture, a major water user is also crucial given its vulnerability to climate change and significant impact on the Middle East's water scarcity. Assessing agriculture's effects on water resources is challenging, but enhancing resilience demands increased engagement with local stakeholders. Choptiany et al. (2019) highlights the need to shift from resilience measurements focused on development actors' needs to empowering farmers to enhance their climate resilience (Choptiany et al. 2019, 28). This shift underscores the importance of local knowledge and practices in managing water resources effectively, reinforcing that strengthening local capacity and autonomy in water management is critical to achieving sustainable agriculture and resilience against climate change in the region.

Innovations in agriculture offer potential for reducing water and power use without sacrificing yield. For example, increasing solar power can significantly reduce power needs while enhancing resilience and creating community-level livelihood opportunities. Additionally, the increasing trend toward indoor, vertical farming creates significant opportunities to increase food production while decreasing water usage. For instance, Bustanica's vertical farm in Dubai, the world's largest, spans 330,000 square feet and aims to cut water usage by 250 million liters annually compared to traditional farming (Peskett 2023). While such advancements are currently more common in wealthier Middle Eastern regions, there is great potential, under the right conditions, for incorporation in more fragile contexts with the assistance of development funding.

Water scarcity and the lack of resource management also pose a significant challenge for ensuring a successful nexus transition between humanitarian and developmental programing in the Middle East. The region's water stress, exacerbated by over-extraction and outdated water systems, presents formidable challenges, especially in nexus transition states, and will require innovative strategies and new, efficient resource management to bridge the gap between short-term and long-term objectives. For example, Jordan is one of the most water-stressed countries globally, with its per capita water availability well below the water scarcity threshold. This has necessitated short-term humanitarian interventions to provide immediate water access to vulnerable populations, particularly refugees from neighboring conflict zones. Transitioning from short-term humanitarian water aid to long-term developmental programing in Jordan has been hindered by resource constraints and the need to address the structural causes of water scarcity, including over-extraction and inefficient water management practices (Schyns et al. 2015; UNICEF Jordan and Economist Impact 2022).

An interviewee, Tim Grieve, highlighted that in Daraa, Syria, just across the Jordan border, "the mining of the scarce water supply for domestic use negatively impacts humanity and makes communities more vulnerable" (Grieve 2023b). Further supported by a REACH (2022) report, he detailed how climatic variations, notably reduced rainfall and the Euphrates River's critically low levels, have deprived over 5 million people in northern and northeast Syria of essential water for drinking and domestic purposes. This situation has led to significant agricultural losses, heightened risks of waterborne diseases, and increased security threats, with a looming escalation in food insecurity and malnutrition anticipated (REACH 2022). Tim Grieve concluded, "In the wider lens of water scarcity in the Middle East, it will be increasingly harder to solve the bigger and longer-term problems like climate change" (Grieve 2023b).

The environment plays a crucial role in exacerbating water scarcity challenges in the Middle East, intersecting with various factors, including the destabilizing effects of climate change and persistent cycles of crises, displacement, and transitioning to and from humanitarian interventions to sustainable development. As the region urgently addresses the need for clean water, adequate sanitation, and hygiene facilities, the complexity of these issues is amplified by environmental influences. Climate change, a global phenomenon, intensifies water scarcity through rising temperatures, erratic rainfall patterns, and prolonged droughts, exacerbating already strained resources in this arid region. These factors increase competition for strained water resources among communities, industries, and agriculture. Furthermore, extreme weather events such as floods and storms directly threaten water infrastructure and sanitation facilities, reshaping the landscape upon which water, sanitation, and hygiene efforts depend. The changing climate reshapes the very landscape upon which WASH efforts are built, necessitating adaptable and durable solutions that incorporate both short-term resilience and long-term sustainability.

Following feedback from our respondents, we argue that water scarcity is among the greatest environmental challenges to addressing WASH needs in the Middle East, exacerbated by climate change and natural and man-made disasters, greatly impacting the transition between the humanitarian-development nexus. Karine Deniel stated in an interview that

> water scarcity and resource management is being overlooked within the region. There is a need to better understand the extent of water scarcity, and look into and analyze the use and overuse of water access, which is also a driver of conflict.
>
> (Deniel 2023)

This was further substantiated during our interview with Jacob Waslander, who stated, "Countries cannot simply pump new water or desalinate water and then bring this water thousands of miles inland, as this brings up major concerns" (Waslander 2023). Water scarcity and the growing need for transporting water from deeper and more distant locations quickly become a multisectoral threat that includes not only environmental concerns but also institutional (i.e., cross-boundary conflicts), technical (i.e., operation and maintenance of infrastructure), financial (i.e., sustainability and dependency), and social concerns (i.e., determine who will distribute water and to whom). It is important to note that environmental challenges remain intertwined and interdependent with all other challenges, from institutional and financial to social and technical. As Waslander succinctly concluded, "Innovations can be more and more impactful, and technical leaps can be taken. However, if the basic public good of water is not there, it becomes difficult to address water needs" (Waslander 2023).

Within the context of the humanitarian-development nexus, the compounding impacts of water scarcity, conflicts, and climate change currently highly underscore the interconnectedness of these factors, amplify vulnerabilities, and complicate both relief and long-term development efforts. Scarce water resources hinder immediate WASH provisions and perpetuate a cycle of health hazards, especially in protracted conflict scenarios. As a survey respondent working for the ICRC stated, "Despite their severity, environmental issues are not widely recognized and accepted. Therefore, there is a large gap to be filled in terms of education and awareness raising." Integrating water scarcity considerations into development programs is thus critical to crafting solutions that address immediate humanitarian needs and long-term resilience.

Solutions to water scarcity in the Middle East require the development of better, more reliable data and greater integration of data with resource management systems grounded in innovation aimed at solving the root causes of water scarcity and climate change. Karine Deniel stated in an interview, "Governments and donors need to work with academic institutions to understand the systems better. Why only do a needs assessment when there is also a need for a deeper analysis of the system" (Deniel 2023). Dialogue around better data is ongoing and was also discussed at length during the 6th annual Arab Water Week in March of 2023, noting the need to understand better the scale and scope of climate change's impact in the region. More research into WASH and climate change is needed. The importance of linking applied research studies in universities and research centers with the WASH sector must be emphasized, especially how this research can be better applied to improving utilities (ACWUA 2023). The conference went on to raise two major challenges facing the Middle East, namely, a need for conducting more studies and projects in the field of water, energy, and food and conducting more studies and finding solutions in the field of reducing the effects of climate change on water resources (ACWUA 2023). Furthermore, the research findings and conclusions must also be communicated and understood by the communities they impact.

Resilience building and sustainability are key to addressing not just the challenges of water scarcity but climate change. However, the current response to water scarcity in the Middle East drastically differs in rich countries, which rely on expensive technologies that bypass solutions and innovations to address the root cause of water scarcity. A recent report from the World Bank, ICRC, and UNICEF (2021) noted that wealthy, oil-rich countries rely heavily on desalination as their primary response to water scarcity. Yet, desalination and wastewater reuse are ineffective and unsustainable solutions for low-income countries in the region. In protracted crisis areas like Yemen, development programs have begun to invest in these high-cost solutions, raising questions about the long-term feasibility of either getting a return on investment or reaching a sustainable equilibrium where local resources can manage the high cost of continued operations (World Bank, ICRC, and UNICEF 2021).

In a crisis, the challenges to resilience and sustainably become exponentially more difficult for governments and institutions to address as "tools available pre-crisis for managing water resources were rapidly lost owing to conflict, weakened institutions, limited mobility, lack of funds, looting of and damage to water resource monitoring networks, and loss of staff" (World Bank, ICRC, and UNICEF 2021, 11–12). Meeting these challenges requires humanitarian actors to design and implement lifesaving water programing with stronger, long-term considerations built in. Humanitarian responders in water scarce regions, particularly in the Middle East and North Africa, often address water needs with borehole drilling as a means of finding and providing water resources to communities in need with little to no coordination or integration with local institutions, capable of managing the long-term implications of groundwater depletion (World Bank, ICRC, and UNICEF 2021). The problem of humanitarian short-sighted drilling for water, damaging or destroying long-term water resources capabilities in the Middle East, has been noted before. One such example was in a World Vision case study in Afghanistan during the 2018 drought that internally displaced at least 3,00,000 people. It was noted that underwater groundwater systems were severely damaged or permanently destroyed through drilling yet came at the cost of decreased long-term resilience and sustainability, increasing the potential of displacement due to drought in the future (Snel and Sorensen 2021; World Vision 2020).

Social Cohesion

Social challenges are often overshadowed by other, more complicated challenges, like the environmental challenges of water scarcity or institutional complexities. However, social challenges are intrinsically interconnected with all other challenges, whereby actors, governments, and beneficiaries grapple with the complex interplay of disrupted social cohesion, the absence of peace-building initiatives, and the imperative to change social behaviors. Yet, effective localization is built upon social cohesion and is essential to the long-term sustainability and resiliency required in effective climate change adaptation. Therefore, overcoming challenges to social cohesion is critical to sustainability and resilience-building efforts to mitigate climate change's current and future impacts. Amidst these challenges, it becomes apparent that addressing WASH needs necessitates innovations in FIETS solutions and a profound reimagining of social dynamics, collaboration, and community engagement.

Geopolitical tensions and conflicts within the Middle East region pose significant social challenges, particularly when viewed in the context of climate change and WASH services. Jacob Waslander noted in our interview that "geopolitical tensions are continuously rising within the Middle East region, most notably related to water and other scarce public goods" (Waslander 2023). The implications of such tensions are widespread and impactful, not the least on the most vulnerable segments of the population. Attacks on water and sanitation services disproportionately affect marginalized groups and not only exacerbate the impacts of climate change but also expand and entrench local, national, and regional conflicts. For instance, children bear the brunt of such attacks, with lifelong consequences. Jeanne Mrad, Permanent Mission of Lebanon to the UN, to the UN Security Council, noted that "in the Middle East, many cities may become uninhabitable before the end of the century." Jeanne Mrad asked, "Is it wars that impact climate change or climate change as a multiplier effect in fuelling civil wars and other armed conflicts?" What is clear is that greater ambition and urgency are needed to stop future conflicts from starting and current conflicts from growing. There is a reinforcing relationship between climate change and conflict, and addressing the root cause of the two will require innovation, leadership, and local buy-in to succeed (United Nations 2023).

International Humanitarian Law (IHL) and international human rights law safeguard children's access to clean water and sanitation. Most notably, Article 24(2) of the UN Convention on the Rights of the Child explicitly recognizes the right of children to access clean drinking water states

> Parties shall pursue full implementation of this right and, in particular, shall take appropriate measures ... To combat disease and malnutrition,

including within the framework of primary health care, through... the provision of adequate nutritious foods and clean drinking-water.

(United Nations 1989)

However, a growing disregard for these rules can be observed in various
armed conflicts, putting the lives and well-being of children at risk. IHL also
protects water and sanitation service providers, classifying them as civilians,
and regards water and sanitation infrastructure as civilian objects vital for
the civilian population's survival. Nevertheless, incidental harm and the mis-
use of infrastructure persist in many conflict-ridden areas. Experts like Tim
Grieve advocate for a conflict-sensitive approach beyond mere risk management to address this complex challenge. Tim Grieve states that "there is a need
to approach conflict sensitivity from not only a risk management perspective
but also to encourage more positive and impactful outcomes and minimize
negative risks alone" (Grieve 2023b). Such approaches aim to yield more
positive and impactful outcomes while minimizing negative risks, making it
essential for the long-term well-being of populations across the Middle East.

These geopolitical tensions also highlight the urgency for diplomatic
efforts and international collaboration to resolve resource-related conflicts as
access to clean water and sanitation remains a fundamental human right, yet,
in an era marked by the growing impacts of climate change and water scarcity,
safeguarding this right is becoming increasingly difficult. Conflicts disrupt the
immediate availability of WASH services and contribute to the overall instability of the region, making the challenges of climate change adaptation and
mitigation efforts even more daunting as humanitarian actors face protracted
fragility with fewer and fewer resources to draw on.

Broader regional geopolitical tensions, along with outright conflict, have a
direct impact on social trust and cohesion. Societal cohesion and trust are the
foundation for building reliable, sustainable, resilient institutions capable of
absorbing shocks during disasters and emergencies. In our discussion about
conflict in the Middle East with Tim Grieve, he argued further that "given the
generations of conflict in the Middle East, social cohesion in society has bro-
ken down, and there is a lack of trust within and between communities, and
between communities and service providers and authorities" (Grieve 2023b).
Without a foundation of social cohesion, building long-term resilient and
sustainable WASH systems will remain challenging, especially in response
to slow-onset disasters like climate change, which complicated transitions
across the HDP nexus. Yet, monitoring and evaluating social cohesion is complicated and less concrete, leading HDP actors to focus on easier programing
with clearer outcomes. Tim Grieve continued that

There is a need for mutually reinforcing, social cohesion in protracted
crisis settings. It is hard to build capacities without social cohesion in
and between communities and service providers. Building resilience in

these systems can help reinforce social cohesion. It remains hard to monitor social metrics and indicators, and less tangible concepts, to ensure resilience.

(Grieve 2023b)

As noted by Grieve, the fractured trust among communities, service providers, and authorities, exacerbated by ongoing conflicts and political instability, hinders effective collaboration, coordination, and sustainable development of WASH systems in the region. Yet, if HDP actors are to succeed in providing sustainable and resilient WASH services, re-engaging with local and national institutions in a way that rebuilds public trust is essential to the long-term outcomes they seek to achieve.

In the Middle East, there is room for more effective behavior change that leads to greater social trust and institutional strengthening. In our interview with Esmail Ibrahim, he emphasized the region's cultural, historical, and educational richness, which fosters a deep appreciation for the sciences. However, he noted a need to further enhance the understanding of climate change's effects and societal consequences (Ibrahim 2023). In our discussion with Denis Heidebroek, it was observed that donors feel challenged in influencing social behavior changes. Despite efforts in hygiene behavior or water usage, these initiatives have not been scaled up to address broader social issues (Heidebroek 2023). This context underlines the need for a multifaceted approach that addresses technical gaps and acknowledges the importance of public trust, transparent governance, and collective efforts in achieving sustainable WASH solutions. Adopting a nexus-thinking, localized approach makes it possible to more effectively meet the varied needs of the Middle East's populations, paving the way for comprehensive strategies to confront the societal challenges posed by climate change in the region.

For these reasons, social challenges are among the most arduous to overcome, especially when addressing the complex issues of climate change and WASH service needs in the Middle East. Social cohesion, a foundational element for effective long-term adaptation and mitigation efforts, has been eroded by generations of regional conflict, creating a climate of mistrust within and between communities. This breakdown in social fabric constitutes a substantial obstacle, highlighting the need for community-led programing and localization. However, despite local expertise, challenges persist due to a lack of effective peacebuilding efforts. This is particularly significant in discussions around the HDP nexus, where collaboration across the humanitarian and development divide is essential to foster resilient and sustainable WASH systems. Recognizing the role of the HDP nexus underscores the need for a harmonious blend of strategies encompassing social dynamics, technical prowess, and operational coordination. Without integrating conflict sensitivity and enhancing cross-sector collaboration, the Middle East will persist in encountering obstacles in addressing complex social challenges. This ultimately hinders progress towards resilient, sustainable WASH solutions capable of withstanding climate shocks and supporting the long-term welfare of its populations.

Local Technical Concerns

Additionally, a greater emphasis on local technical capabilities is critical to long-term regional resilience building. The Middle East's pursuit of addressing technical challenges surrounding WASH needs, including efficient water treatment, sanitation systems, and sustainable hygiene practices, is burdened by the complex interplay of technical gaps, a lack of skills and capacity, and the need for community-led programing. Forced migration is directly linked to the gaps that local, marginalized groups face in adapting to climate-related shocks. As the region grapples with the interdependencies of these challenges, it becomes apparent that addressing WASH needs necessitates innovative technical solutions and a profound reimagining of collaboration and capacity building, including low-cost appropriate solutions. Filling these technical gaps represents an opportunity to leverage local assets and insights from local decision-makers fully.

The widespread lack of skills for addressing WASH needs in the Middle East significantly challenges the region's capacity to effectively plan, implement, manage, and secure funding for WASH programs and projects. This point was noted in our discussion with Khaldon Khashman, who stated that

> There is a general lack of skills in the humanitarian and the WASH sector. Building the capacity of local people should be a major new focus within both. Local experts lack skills and experience in how utilities are managed, quality control, and water safety plans. There is a need for skills more generally among operators, engineers, and the utilities sub-sector. There are great technical people within the region, but they lack these specific skills.
>
> (Khashman 2023)

These technical gaps extend beyond the humanitarian and WASH sectors and were also identified at the governmental and institutional levels in our interview with Jean Christophe Barbiche, who stated that "there is a lack of technical capacity by the government and the population to set up, maintain and financially sustain more expensive impactful projects (i.e., bigger pumps and deeper boreholes)" (Barbiche 2023). Barbiche described how most local and national governments lack the technical capacity and financial abilities to install and maintain larger WASH infrastructure. If the financial hurdles are overcome, attracting funding for more innovative WASH projects that address broader and deeper WASH needs will become easier. With this funding, technical challenges and gaps in expertise among government staff and institutions responsible for maintaining such projects will decrease, resulting in stronger, more sustainable services.

These technical skill deficits are critical to upgrading the current WASH infrastructure with critical climate-specific sustainability and resiliency redundancies while preparing for anticipated shocks and increased service needs due to a rapidly growing population. This poses a risk to governments in fragile states who wish to transition into development programing, whereby donors would be less inclined to fund innovative projects if the government cannot complete the project's handover and sustain the operation and maintenance for more advanced and impactful WASH systems. Insufficient technical skills among professionals and practitioners within the WASH sector ultimately hinder developing and applying innovative and context-specific solutions. This lack of expertise can result in inadequate infrastructure design, suboptimal maintenance practices, and an inability to address complex water and sanitation issues, such as pollution control and wastewater treatment. Additionally, the absence of specialized skills in areas like water resource management, hydrology, and environmental engineering further complicates efforts to address the region's unique challenges, particularly the impact of water scarcity and climate change.

Overall, the shortage of technical skills undermines countries in the Middle East's ability to ensure access to clean water, proper sanitation, and hygienic facilities for their population. This challenge raises the need for training programs in various WASH aspects for those working within utilities throughout the region. During our interview with Farah Al-Basha, she noted that critical "technical knowledge could come from other sources within the Middle East" (Al-Basha 2023). As such, even if local WASH-specific skills are missing, WASH-adjacent capabilities may be necessary to fill and meet needs. Furthermore, if the necessary capacities and capacity builders are not available within the region, international experts must prioritize the training of local WASH capacity both in-country and within the region.

An increasing number of humanitarian organizations are leaning on local capabilities to deliver aid for their humanitarian missions, either by self-sustaining their interventions or by autonomously spearheading initiatives to address the fundamental requirements of affected communities. Nevertheless, even in cases where local skills are present and underpinned by robust academic backgrounds, what is frequently lacking is the humanitarian perspective and the firsthand experience that would enable them to operate at their maximum potential while transitioning more senior roles to local professionals within the humanitarian field. One of the key developments in the WASH sector in recent years has been the establishment of one of the first master's programs aimed at developing local capacities in WASH. The program, through the German-Jordan University in partners with Action Against Hunger (ACF), Bioforce, UNICEF, the GWC, and other humanitarian organizations, aims to build the capacities of local WASH specialists through a humanitarian lens and help them provide the necessary human capital needed to deal with the region's response to humanitarian crises while increasing the interest of future potential specialists to enter the field (GJU 2020).

Financial Realities

Navigating complex and inflexible financial structures poses a significant challenge for HDP actors and institutions as these frameworks impede effective collaboration and coordination, obstruct transitions within the HDP nexus, and undermine sustainability and resilience efforts. This scenario becomes more critical in the Middle East, where the interplay of water scarcity, climate change, and prolonged crises exacerbates financial vulnerabilities, threatening the effective transition across the HDP nexus and the achievement of both immediate- and long-term WASH goals. As was argued in our interview with Farah Al-Basha, who stated that "Climate change is a major issue for which financial stability and sustainability are key to introducing mitigation measures. Additionally, the region's political instability adds a multiplier effect on the financial vulnerability, which consequently impacts WASH services" (Al-Basha 2023). Noting the interconnected nature of climate change, sustainability, political instability, financial vulnerability, and WASH services, we argue here that these challenges are rooted in low levels of investment in building and fostering the capacity of the WASH sector, financial dependency, and poor financial sustainability, resulting in inadequate technical and institutional expertise, weak accountability, and fragile financial stability. Additionally, climate change's increasing toll underscores the urgency of introducing adaptation and mitigation measures, the viability of which rests heavily on financial and economic stability. Nevertheless, the challenge becomes increasingly intricate in light of geopolitical instability, which amplifies financial vulnerabilities, thereby directly impeding the region's capacity to address critical weaknesses in current WASH services. However, these challenges in the Middle East are further complicated by regional instability, which acts as a compounding multiplier on social, technical, and financial vulnerabilities, impacting the region's ability to address climate-related WASH challenges sustainably and resiliently.

In fragile states, funding and prioritization are significant issues. With so little funds to draw on, it is hard to prioritize long-term climate change adaptation and mitigation efforts over other pressing short-term priorities. With the lack of financial sustainability for humanitarian and development programs among governments throughout the Middle East, climate and WASH programing requires international financial dependency to cover the implementation costs. During our interview with Farah Al-Basha, she argued further that,

On one side, there is a reluctance by the government to take over with no return for investment from their side. On the side of the actors, there is an inability to hand over things to the government because the government doesn't have the financial capacity to sustain and lead these efforts. There are behavioral issues on both sides, and the government is dependent on funding from the international community.

(Al-Basha 2023)

WASH interventions and systems require increased ownership and funding from governments and international actors and an increased focus on financial realities that prevent successful ownership transfers to local and national institutional bodies.

Addressing the issue of ownership requires a nexus mindset by all parties during each step, from design to implementation. A survey respondent with the ICRC noted that

> While the elements presented cannot be seen in isolation, the main challenge of water supply systems, for example, in the Middle East, is that they depend on state subsidies by design. In other words, during their design phase, development actors did not consider the 'nexus,' i.e., the systems' resilience(s) in case of protracted crisis.

With proper ownership and control over the operation and maintenance of WASH systems and sufficient investment and subsidies on the government side, WASH systems can be sustained. Although international actors can fill that role without government support during any initial humanitarian crises and emergencies, in the end, interim and long-term ownership belongs to local and national institutional bodies. This highlights the challenge of financial dependency across the HDP nexus as programs transition towards government ownership. This all entails that there needs to be more nexus thinking by all actors, including donors and governments, notably during the design phase of projects, wherein development actors need to consider closely the nexus transition in terms of transferring the ownership and financial burden to governments for addressing WASH needs to ensure the systems' resilience, particularly in cases of protracted crises.

This predicament stems from a complex interplay of factors, including government reluctance to assume full responsibility without corresponding adequate returns on investment and actors' inability to transfer programs due to the government's limited financial capacity. Governments, at times, are hesitant to take full responsibility unless there is a clear correlation between their investment and the expected outcomes. This dynamics underscores the need for a comprehensive reevaluation of funding models and partnerships to ensure the continuity and resilience of vital programs, notably in the transition toward development. Achieving greater financial self-reliance for governments and promoting sustainable development calls for innovative approaches that address the behavioral challenges on both sides and foster a shared commitment to long-term success, which is required for the larger issues of water scarcity and climate change to be addressed.

However, a lack of funding is a concern not just at the local level but also at the global level. Only five years ago, the humanitarian WASH response was assessed to be 47% funded. Post-COVID-19, that number is likely lower

(Global WASH Cluster 2019). The impacts of low funding are severe. The GWC stated that

> The capacity of the WASH sector to lead, coordinate and deliver humanitarian assistance is a key driver to the quality and impact of the response. However, a low level of investment in building and fostering this capacity has results in inadequate technical expertise and weak accountability. The need for a versatile WASH sector that operates as part of an improved response model has become increasingly central to the delivery of targeted interventions in rapid onset and acute emergencies, and to durable and sustainable solutions in complex and protracted emergencies... The response capacity of the WASH sector is of paramount importance in balancing the broad range of programmatic, coordination and leadership considerations while maintaining the humanitarian imperative and building linkages to longer-term development.
>
> (Global WASH Cluster 2019, 6–7)

Greater investment is needed to meet the increasing WASH needs in fragile states and crisis areas. Specifically, limited funding for WASH is a major challenge in the Middle East. Denis Heidebroek, describing humanitarian response plans (HRP) in the Middle East, stated,

> The available tools you have out there are underfunded, and the wash sector therewithin. It is not uncommon to have the HRP through the Middle East funded at 10% or even less for the WASH sector according to its needs.
>
> (Heidebroek 2023)

In addressing climate change specifically, Fabrizio Carboni, ICRC Regional Director for Near and Middle East, argued in an interview on Al Jazeera in June 2023, that this is a regional trend and that a lack of funding for humanitarian aid in the Middle East is hurting the communities most in need, with Syria, Lebanon, Yemen, Iraq, and Palestine among the worst affected (Al Jazeera English 2023). Carboni argued further that

> There is a climate crisis, and funding is decreasing. We want traditional donors to do more, but we also need other actors to come. The burden needs to be shared by more actors than the ones who leading humanitarian action over the last decade... [The climate crisis] is a public health crisis, because you don't have water, it's economic, because if your harvest are affected and you don't have what you need for food and to kickstart the economy. You cannot isolate the climate crisis from all aspects of life. That is what makes it very urgent for, not only humanitarian actors, but for the whole international community observing those situations of conflict and

instability to be serious about it and to channel the necessary funding and technology to those areas affected by climate.

(Al Jazeera English 2023)

The challenges and consequences of climate change are much larger than local, national, and regional concerns, requiring international and global commitment and funding.

Among the funding provided within the WASH sector, there are concerns about the limited availability of donor grants allocated to climate-sensitive and green approaches. In the absence of funding for climate-sensitive approaches as a sector in and of itself, humanitarian and development organizations, which rely heavily on donor funding, are forced to integrate climate approaches as a cross-cutting theme through all other interventions. During our interview with Jean Barbiche, he stated that

> now when we do WASH projects and even other projects across other sectors, we try to integrate climate sensitive programming, keeping in mind environmental degradation, and other approaches such as water/rain capture from roofs and soil, and the reuse of water from organic waste.
>
> (Barbiche 2023)

In all, donors need to play a far greater role in making it easier for conflict-affected countries and communities to access financing for climate adaptation. There is a need to ensure that populations affected by the combined impact of conflict, climate change, and environmental degradation receive the support they need.

There is also a need to change how donors, actors, and governments think and act in relation to climate change. Denis Heidebroek explained that "some donors, such as ECHO [European Civil Protection and Humanitarian Aid Operations], have most recently introduced mandatory minimum environmental requirements, which consider resilience programming" (Heidebroek 2023). This is a practice that Heidebroek recommends and should be adopted by other donors and actors. Discussing this point further, Jean Barbiche argued that,

> There is a need for a new emerging strategy that integrates environmental and resilience into all programming. In the Middle East, resilience programming is less of a focus than in other regions like Africa, where resilience is integrated into all projects. Resilience is everywhere.
>
> (Barbiche 2023)

Regarding climate change, discussions have predominantly focused on the amount of funding available, with less attention on ensuring that funded climate resilience projects are adequate, robust, and effective – key to attracting alternative funding streams (UNICEF and GWP 2022). Karine Deniel explained that "on the donor-facing side, donors also need to understand better humanitarian action, and how resilience thinking in relation to climate change can be better incorporated into their work" (Deniel 2023).

Sustainability and Resilience Thinking

Sustainability and resilience are age-old terms that are nonetheless emerging as key concepts for their potential as foundations for climate change adaptation strategies. The necessary change is a fundamental shift in how we think about and plan for the future challenges of climate change. Sustainability and resilience frameworks can thus become a shared vocabulary and lens through which to view current problems and plan much-needed adaptations. Sustainable thinking emphasizes the importance of innovation in addressing WASH challenges and achieving desired resilience outcomes. This approach includes sustainable water supply systems, waste management, and water reuse to meet immediate and future needs, among many others.

Two of the greatest pressures from climate change, particularly in the Middle East, are water scarcity, which we have attempted to address here, and growing concerns around energy usage, which is vital to sustainably providing water services. Addressing energy battle necks and creating secondary, more localized power sources is a critical first step toward building more sustainable and resilient water services. UNICEF and the GWP have both highlighted the need to lower the WASH sector's carbon footprint by not only increasing the efficiency of water energy use but also relying more heavily on green systems like solar power and other forms of renewable energy (UNICEF 2020, 6; UNICEF and GWP 2022, 5). UNICEF and the GWP went on to note that aside from the possibilities of solar power as an alternate, localized form of energy, other potential energy sources included "recovering biogas from wastewater and fecal sludge and generating energy from waste [and] the reuse of wastewater using renewable energy [are other] important opportunit[ies] to reduce emissions, ensure adequate treatment of wastewater and reduce the demand for freshwater" (UNICEF and GWP 2022, 6). This can help reduce greenhouse gases, a key cause of rising temperatures globally, and also help produce new skills development and livelihood opportunities as new technologies and their maintenance are integrated into local and national systems. UNICEF and the GWP's 2022 report connected the need for greater WASH resilience with the potential for long-term economic outcomes, noting that "Acting now provides an opportunity for policymakers and service providers to rethink access to basic services and contribute to a circular economy and green growth, thereby supporting job creation and the development of skills" (UNICEF and GWP 2022, 5). The sooner action is taken by humanitarian, development, and government actors, the greater the impact new policies not only will have on increasing the resilience and livelihood opportunities for individuals, families, and communities living in fragile contexts across the Middle East region but will also go a long way toward reducing the potential for further conflict and forced migration (UNICEF and GWP 2022).

There is still time to focus on and build more resilient water systems, but the pressure faced by marginalized groups in increasingly fragile contexts in the Middle East is growing. If funded accordingly, foundational longer-term change can start first, focusing on sustainability and resilience building that supports longer-term interventions regardless of the short-term or long-term nature of humanitarian and development actors working in fragile contexts. During an interview, Andy Bastable stated, "Both actors should focus more on establishing sustainable water supplies and new innovations such as solid waste management and water reuse. Such interventions address immediate and longer-term needs, all in one go" (Bastable 2023). With careful consideration, in the right contexts, many water-related humanitarian crises can be addressed in the short run in a way that also builds long-run increases in water resources, thus building resilience and sustainability and reducing the need for future humanitarian assistance. We highlighted a similar case in Afghanistan in a previous publication where macro-catchments were used to slow precipitation runoff during a drought, which not only created long-term usable reservoirs for local use but also began refilling long-dormant underground water aqueducts, significantly increasing scarce water resources (Snel and Sorensen 2021; World Vision 2020).

Yet, these solutions are not the norm, and stronger, more purposeful resource management is needed. In our discussion with Jean Christophe Barbiche, he noted that there has been a significant challenge for farmers and agriculture in places like Iraq and Libya, where there have been several years of ongoing drought. Barbiche went on to highlight an overreliance on outdated systems and solutions that were not and are not suited to meet the unique challenges of climate change, stating that,

> In Iraq, more and more boreholes are being installed, so the impact is not yet visible in the short-term to the general population. However, farmers are seeing the negative impacts of climate change firsthand hand the present day. Same as in Libya, there are networks of irrigation canals bringing water from rivers to areas where there is no water but instead land to cultivate. The question must be raised as to how long this can be maintained. The irrigation system, as it was before and is today, is not sustainable.
>
> (Barbiche 2023)

The water systems of the past, though providing water now, may not be resilient to climate-related changes of the future.

These same observations were also brought up in several examples given during an interview with Omar El Hattab, who noted that in the Sahel region of the Horn of Africa, failed rainy seasons have led to a significant increase in borehole drilling originally at depths of 200–300 meters, but as ground waters are being overused they are now looking at increasing the depth to 500–600 meters (El Hattab 2023). Yet, better water management and greater long-term foresight are needed as he showed in an example nearby in Ethiopia, arguing that

> Ethiopia receives nearly 1 billion cubic meters of precipitation annually. Yet certain regions of Ethiopia, like afar, are challenged with drought. Well, what does that tell you? It tells you that not adequate investments have gone into surface water management. Because they are having floods in one region and drought in other regions. This calls for immediate attention to investing in the long-term development of water resources.
>
> (El Hattab 2023)

Though not a Middle Eastern example, they highlight the need for a greater understanding of the impact climate change has on increasing droughts and floods and, critically, the need for greater long-term investment in water resource management, built on principles of resilience and sustainability, even in, and especially in, fragile states. This requires humanitarian actors to think longer term and development actors to find ways to reduce risk and get involved sooner.

At its core, sustainable interventions ensure better management of water resources and, when designed and implemented well, create cost-effective and environmentally friendly projects. In 2016, Howard et al. stated, "Sustainability is not simply an assessment of operational functionality at any one point in time, but needs to be considered as a wider set of institutional, financial, and environmental issues" (Howard et al. 2016, 257). Doing this requires bringing together humanitarian and development actors' strengths in fragile states. During our interview with Andy Bastable, he argued,

> Some would say that climate change isn't changing the role of humanitarian and development response, as all responses need to be delivered with sustainability as the main focus. Sustainability should not be the focus of only long-term actors, and not short-term actors, but both.
>
> (Bastable 2023)

The Middle East is facing a compounding future where climate change-related water scarcity, compounded by population growth and greater urbanization, is creating a reality where fragility and marginalization are only anticipated to increase. Conflict and forced migration will only further exacerbate the strenuous equilibrium in the region, and the likelihood of long-term, persistent, protracted crises is high. Success will require not just bridging between the humanitarian and development silos but integration and merging between silos, typically split in two by complex funding streams.

The evidence is becoming more clear that climate change is currently having a significant impact in the Middle East as water becomes more scarce, leading to increased conflict and forced migration, particularly in marginalized groups who often lack the skill sets or resources necessary to adapt adequately. The slow onset nature of climate change makes it easy for governments, institutions, and humanitarian and development actors to push off responsibility into the future. Yet the consequences of climate change are happening now as conflict and crisis become more protracted in length, complicating humanitarian and development transitions in these fragile states. Across the Middle East regions, water and sanitation systems and infrastructure lack vital resilience to droughts and floods, which are common effects of climate change, while often relying on strained and unsustainable energy systems ill-prepared for the coming pressures of an increasing and more urbanized population. This puts water and sanitation systems at risk of catastrophic failure, increasing the potential for conflict over scarce water supplies and increased health risks as more people, pushed to the margins, have to rely on contaminated or unsecured water sources. However, innovations in technology and management systems can serve as potential solutions to current and future climate-related challenges, but only if they rely on resilience and sustainable frameworks. Howard et al. noted that some of the simplest, most effective solutions to the growing challenges of climate change involve building resilience into water plans and risk management systems, which is an easy first step for even low-income countries (Howard et al. 2016).

Localization is vitally important to successful and sustainable resilience building. Local leaders and institutions must be consulted and lead in coordination if interventions are to become sustainable (Baxter et al. 2022; Satterthwaite et al. 2020). This is especially true in humanitarian contexts and fragile states, where local institutions may be weak or appear absent altogether. Yet this provides an opportunity to build from scratch or rebuild and change institutions that are shocked or damaged after a crisis. Utilizing assets instead of deficit-based frameworks allows humanitarian and development actors to refocus on local capabilities that communities still have and can bring to the table, even when, at first glance, they are shattered beyond repair (Winterford, Rhodes, and Dureau 2023). This is vital for sustainable resilience building and developing transitional bridges between humanitarian and development actors, providing much-needed exit strategies for over-stretched humanitarian funds. Urbanization presents a significant challenge as growing populations, notably marginalized groups least equipped to handle climate change, migrate to cities and face unique challenges in securing livelihoods. In a recent publication, Satterthwaite et al. stated that urban communities have rapidly grown informal settlements that house as much as a fifth of the world's urban population (Satterthwaite et al. 2020). These informal settings are particularly at risk of the ongoing threat of climate change. They will require a significant increase in local commitment and cooperation if climate resilience programing around water, sanitation, and health will make an impact. This will require a critical rise in internal and external funding and an increase in effective community-driven programing to succeed. Without this, water resources will continue to become more scarce, contaminated, and contested, leading to conflict, further migration, and adverse health outcomes.

Yet, targeted social protection systems, resilience building, and more effective resource management can reduce conflict, increase health outcomes, and develop new relevant skills, leading to greater economic and livelihood outcomes for those most impacted by climate change (Malerba 2021). Resilient and sustainable approaches to water, sanitation, and energy management increase marginalized groups' ability to healthily and consistently obtain much-needed resources, even in fragile states. Yet when funded and appropriately designed, climate-related resilience building also grows marketable skill sets, increasing livelihoods and leading to greater economic outcomes at both the local and national levels.

The discourse around climate change continues to increase as climate change's long-term impact on countries, economies, and societies becomes clearer. Debates are become more strategically focused on what can be and should be done, with a growing focus on a "well-conceived integrated water resource management [which] should be the overall water governance framework within which climate change adaptation should take place" (Batchelor, Smits, and James 2011, 19). Still pertinent, a decade ago, Batchelor, Smits, and James championed a "learning-by-doing" strategy for building water and sanitation resilience in drought and flood-prone areas. They underlined the crucial role of local communities in identifying sustainable resilience adaptations for ongoing and future water and sanitation challenges in climate-affected regions (Batchelor, Smits, and James 2011). Community-led programing was and continues to be the foundation of effective, long-term climate change response. Yet building sustainable resilience into existing water and sanitation systems in the Middle East will require significant national and international investment if climate change's current and future impact in the region is to be mitigated.

3 Resilience Building and Institutional Concerns

The Middle East faces intricate geopolitical challenges exacerbated by climate change, political volatility, resource scarcity, and fragile institutional frameworks. Our interviewees suggest that the predominant challenges facing climate change response in the Middle East were institutional challenges. Ramifications from growing institutional challenges shape the region's response to climate change, crises, and the broader transition across the HDP nexus. Weaknesses in governance and institutional frameworks emerge as substantial impediments significantly impacting the resilience and sustainability of essential climate-related systems. Interviewees highlighted the core institutional challenges as political instability, including a lack of transparency; geopolitics; poor institutional arrangements, coordination, and shortfalls; a lack of institutional sustainability; and institutional actions, including policy changes, regulations, access constraints, aspects of inclusion, and resource management.

Here, we briefly examine the institutional challenges faced in the Middle East, including issues with institutional arrangements, and argue for more robust, resilient institutions. Our interviewees highlighted deficiencies, emphasizing weak institutions' impact on water resource management, environmental protection, and the broader HDP nexus transition, discussed later in Chapter 4. These regulatory gaps hamper water resource management in the face of changing climate conditions, leading to further political instability. We explain further how this lack of institutional planning and positions on climate change leads to an absence of a clear vision, roles, and responsibilities, resulting in a reactive rather than proactive approach to climate change disasters.

Institutional shortcomings in security and access constraints significantly impact the provision of essential services, most notably in regions affected by conflict. Conflict and fragility hinder donors' and implementors' ability to deliver timely and effective solutions, resulting in delayed responses, extended project timelines, and compromised outcomes. Issues of equal rights and inclusion further complicate climate change efforts to reach the most vulnerable populations. Challenges stemming from weak institutions and the lack of policies that ensure equitable access become crucial to future climate

DOI: 10.4324/9781003436706-3

mitigation efforts, especially for vulnerable populations. The disproportionate impact on groups like women and girls and people living with disabilities, which, coupled with existing disparities amplified by climate-induced stresses, necessitates a commitment to more gender-sensitive and inclusive approaches. These institutional challenges, if unaddressed, threaten both service availability and societal stability. Our perspective highlights the roles of HDP actors, emphasizing the challenges governments and institutions face in ensuring WASH rights amidst climate-related crises in the Middle East.

Political Dynamics and Governance

Amidst the repercussions of escalating climate change, Middle Eastern countries contend with a persistent challenge of governmental and political volatility, posing a significant threat to achieving resilient and sustainable WASH outcomes in the region. The many countries within the Middle East find themselves among the most vulnerable globally to the adverse effects of climate change, which not only jeopardizes the physical environment and societal fabric but also undermines the reliability, resilience, and sustainability of governmental institutions (Abel et al. 2019; Alaaldin 2022; Baghdadi 2022; Waha et al. 2017). Political volatility in the Middle East is deeply rooted in historical, sociocultural, and economic factors and new pressures like climate change, which create a volatile environment that challenges the establishment of stable governance. The region has experienced a series of conflicts and geopolitical tensions, often linked to struggles for power and, in recent years, in response to climate-related causes (Abel et al. 2019; Alaaldin 2022). According to a report by Justin Gengler with the Carnegie Endowment for International Peace, historical grievances, economic vulnerabilities, the manipulation of identity politics, and sectarianism by authoritarian actors are other factors that contributed to the ongoing political turbulence in the Middle East (Gengler 2016; Rizkallah et al. 2019). Ultimately, these constant fluctuations undermine efforts to build enduring political institutions.

These factors are further complicated because a significant portion of the region's wealth is derived from oil resources, leading to economic vulnerabilities tied to fluctuations in global oil prices. The Brookings Institution highlights that the heavy dependence on oil revenue can create economic instability, exposing these nations to the volatile nature of the global energy market (International Monetary Fund 2016). This economic fragility, coupled with high youth unemployment rates, creates a disenfranchised population susceptible to political unrest. The interplay of historical grievances, sectarian tensions, and economic vulnerabilities collectively contributes to the persistent political volatility in the Middle East, shaping the region's geopolitical landscape.

Climate change in the Middle East also exacerbates conflicts and instability through various channels. The region faces climate change-induced pressures, such as wars, droughts, and famines, leading to further economic decline and social corrosion (Abel et al. 2019; Alaaldin 2022; Baxter et al. 2022; Malerba 2021; Norwegian Red Cross 2019; UNICEF 2020). Akther and Alam (2020) argue that falling agricultural production due to droughts and desertification sparks conflicts, fostering tensions and violence among ethnically diverse groups. While climate change is a contributing factor, geopolitical influences play a primary role, with climate issues exploited for war and propaganda. Economic challenges, including slow-moving economies and high unemployment, contribute to unrest and demonstrations – internal factors like social inequality, sectarian conflict, and resource manipulation. Climate catastrophes amplify these challenges, pushing populations to rebel against deep-rooted inequalities and global resource exploitation, ultimately posing a global threat to economic and political stability (Akther and Alam 2020).

This is important to keep in mind since water has the potential to be a more vital geopolitical resource than oil, with demand set to rise more than 50% by 2030 (Ghali 2022). The intertwined impact of weak institutions and political volatility creates an environment where essential services crucial for water security, public health, and well-being struggle to attain the necessary stability, resilience, and efficiency required to meet the pressing demands in the region. Political volatility exacerbates these challenges, fostering inconsistent resource allocation, fragmented coordination, and unreliable service delivery. Consequently, sustainable WASH solutions encounter barriers as programs attempt to transition across the HDP nexus.

Both survey and interview respondents highlighted a reciprocal relationship in the region whereby political volatility acts as a catalyst for the emergence of weak institutions, while these institutions, in turn, perpetuate political volatility. The tumultuous political landscape breeds uncertainty, preventing the establishment of stable governance structures and impeding the development of robust institutional frameworks. This cycle hampers the region's ability to build and maintain institutions capable of addressing the complex challenges posed by climate change in the region. This underscores the imperative for enhanced governance and stability and highlights the critical need to address the intersection of climate change, security, and development challenges in the Middle East (Congressional Research Service 2023). From our research, surveys, and interviews, it is evident that fostering political stability and strengthening institutions is not only a prerequisite for sustainable WASH solutions but is pivotal in building resilience against the intricate web of challenges confronting the Middle East as the region seeks to mitigate and address the current and future challenges of climate change. Yet complex political dynamics can breed uncertainty, disrupting the establishment and functioning of stable governance and robust institutional frameworks. As a result, essential services vital for water security, public health, and well-being face challenges in achieving the necessary stability, resilience, and efficiency. These factors directly impact the implementation and sustainability of WASH solutions, underscoring the crucial need for enhanced governance and political stability to improve climate-related outcomes in the region.

Institutional Arrangements and the WASH Sector

Institutional arrangement generally refers to how organizations, structures, and systems are designed and organized to achieve specific goals or functions within a particular context. In the context of the SWA partnership, which is a global multi-stakeholder platform working towards universal access to clean water and adequate sanitation, an institutional arrangement refers to the organizational structures and mechanisms put in place to facilitate collaboration and coordination among various stakeholders involved in the WASH sector (Sanitation and Water for All 2020b). In this framework, the five key building blocks of a well-functioning WASH sector are sector policy and strategy, planning monitoring and review, capacity development, and institutional arrangements (Sanitation and Water for All 2020a, 2020b). The SWA identifies the building block of Institutional Arrangements as having:

1 The identification and allocation of institutional roles and responsibilities, including decentralization commitments.
2 Country-driven and inclusive coordination mechanisms that allow for participation of a broad range of stakeholders in dialogue, communication, and identification of mutual interest around service delivery and sector learning.
3 Legal and regulatory frameworks to underpin the desired targets and reinforce roles and allocation of resources (Sanitation and Water for All 2020b).

However, in the Middle East, there are many challenges to the building block of institutional arrangements within the context of WASH. The severity of these challenges was noted during an interview with Omar El Hattab, who stated that "poor institutional arrangements, coordination, and resilience pose a major threat, if not the greatest threat, to addressing WASH needs in the Middle East" (El Hattab 2023). First, the identification and allocation of institutional roles and responsibilities, particularly in the context of decentralization commitments, face hurdles due to political complexities and centralized governance structures in some Middle Eastern countries. Achieving meaningful decentralization requires overcoming political resistance and ensuring that local authorities have the capacity and resources to manage water and sanitation services effectively while not alienating vulnerable communities (RBAS 2013; UN ESCWA 2019). Second, country-driven and inclusive coordination mechanisms also encounter challenges in the Middle East. Political tensions and limited space for inclusive dialogue can impede the establishment of coordination mechanisms that genuinely involve a broad range of stakeholders. In some cases, conflicts and geopolitical issues sometimes hinder collaborative efforts, making it difficult to create platforms that facilitate effective communication and mutual understanding among regional stakeholders (World

Bank 2018). Lastly, legal and regulatory frameworks, crucial for underpinning desired targets and resource allocation, face challenges rooted in legal complexities and variations across Middle Eastern countries. Inadequate and restrictive regulatory frameworks can hinder the achievement of WASH targets and the effective allocation of resources. The lack of harmonization in legal and regulatory approaches across the region is not helping to create an enabling environment that supports the sector's development and sustainability (OECD 2018, 2021).

Despite ongoing efforts to improve institutions, there remains a significant gap between mere strength and true resilience. El Hattab emphasized this critical issue in the Middle East, where institutional arrangements are generally sub-optimal, lack resilience, and substantially impede efforts to address WASH needs in the Middle East. He pointed out that although efforts are made to strengthen these institutions, strength does not necessarily equate to resilience. A resilient WASH sector, like other sectors, should be capable of withstanding shocks such as natural disasters or the impacts of climate change. Currently, the institutional frameworks in the region require more resiliency to ensure effective and sustainable operation in the face of such challenges (El Hattab 2023). This was also reinforced during our interview with Esmaeil Ibrahim, who stated,

> Many governments throughout the Middle East do not have the institutional setup and plans in place to prepare themselves to be reactive or proactive to situations and hold themselves accountable to the impact of climate change and disasters that can lead to humanitarian crises.
>
> (Ibrahim 2023)

This emphasizes the need for local and national leadership across the Middle East to address institutional weaknesses in a way that builds resilience and the ability to act proactively to address future climate change shocks.

A 2021 report by the World Bank, ICRC, and UNICEF pointed out that in 2010, there was nearly universal access to water supply in the Middle East. Yet, over the last decade, conflict, political instability, and climate change have significantly reduced that coverage. This is partly because universal coverage can hide significant institutional and sectoral weaknesses, creating the illusion that the WASH sector is strong and resilient at large. (World Bank, ICRC, and UNICEF 2021). These weaknesses in resilience and sustainability, as highlighted by El Hattab, require significant funding to upgrade. The lack of funding in the Middle East to address WASH needs is a major concern. During several interviews, we discussed broader institutional weaknesses, weak tax structures, and a general lack of will to address climate change and its anticipated consequences on water scarcity and WASH services. Andy Bastable stated, "Authorities within the central government are not pushing to contribute sufficient money to addressing WASH outcomes and climate change. Meanwhile, their revenue systems are antiquated" (Bastable 2023). This point was corroborated by Anna Rubert when she indicated that there are also added aspects of weak institutional structure and lack of optimal capacity that "result in financial payments or tax systems not being adhered to, which add to many additional complexities" (Rubert 2023). Finally, a general lack of will across regional institutional decision-makers impedes the development of new financial structures capable of creating long-term climate-focused sustainability in WASH services. Summarizing this, Farah Al-Basha argued that "overall, there is a reluctance to take financial issues seriously on the side of improving WASH outcomes. Water issues are a security issue, which is one reason why water tariffs are not taking place. It is too political" (Al-Basha 2023).

Yet, there is great potential for new innovative funding streams that utilize public-private partnerships to bridge deficits in financial resources. Research conducted by Saadeh (2019) in the field of solid waste management suggests that public-private partnerships can significantly contribute to the sustainability and resilience of environmental management systems, indicating broader applicability to sectors such as WASH. However, this approach necessitates governmental support, legal and regulatory frameworks, and strategic planning to ensure the success and sustainability of public-private partnership initiatives (Saadeh, Al-Khatib, and Kontogianni 2019). However, these funding structures require stable and effective tax revenue to succeed. Political will from local and national institutions is required to develop necessary tax structures and complex public-private partnership agreements that incentivize private investment while meeting context-specific local needs. Actualizing this potential will require a mindset shift towards the uncomfortable recognition that climate change and water scarcity will have a larger negative impact across the Middle East without a significant course correction.

Regional institutions must also help subsidize the costs of addressing the region's climate-related water scarcity and other WASH needs. Blended financial systems utilize a balance between subsidies, taxes, and tax breaks as tools to incentives private partners to meet WASH needs while simultaneously reducing the financial responsibility that poor and marginalized populations have to bear for the operation and maintenance of resilient WASH systems (Fonseca and Pories 2017; Snel and Sorensen 2021). Anna Rubert highlighted this complex balance further by noting that "high[ly] subsidized systems by the ministries are resulting in low levels of payment and thus the reducing the capacity to cover base operation and maintenance cost, let alone any new or future investment in infrastructure, expansion, or necessary staffing" (Rubert 2023). There is a need for awareness raising within communities to encourage payment of taxes or fees for the maintenance and operation of WASH systems. A survey respondent working in the region for ACF stated, "If the population were to better understand the value and the limitedness of water resources through awareness raising, fees would be easier to collect and technical sustainability would be much easier to attain." Local fees are ultimately required to maintain and improve water system resilience and sustainability. The water crisis in the region could be reduced if local institutions responsible for providing WASH services could more efficiently recruit and retain qualified personnel while also effectively managing, operating, and upgrading WASH systems to address emerging challenges associated with climate change.

Our interviewees indicated the greatest challenges to achieving WASH goals throughout the region are at the intersection of weak institutional arrangements, poor governance, and increasing climate variability. Anders Jägerskog stated during our interview,

> The combination of governance and environment will pose the greatest threat because of the setup of institutions and the WASH sector. With increased climate change and variability, the institutional setup of the sector will face the most challenges (including utilities to cover costs, heavy subsidies, etc.)
>
> (Jagerskog 2023)

The point here is that even as local and national institutions seek to address climate change's current and future challenges, the inherent weaknesses in governance policy and institutional arrangements lead to short-sighted solutions that, in many cases, reduce the overall resilience of WASH services.

This deficiency brought about by an inadequate institutional setup hampers effective planning, coordination, and implementation of WASH initiatives. It inhibits the region's capacity to adapt to changing climate conditions and sustainably manage water resources. This emphasizes the importance of resilient institutions, harmonized coordination, and accountability in ensuring the delivery of essential WASH services amidst evolving challenges such as climate change and development transitions. However, in the absence of an established institutional set-up and framework, governments in the region need to be more accountable to carry out their many roles effectively. During our interview with Denis Vanhontegem, he stated that "governance issues within water authorities and institutions need to act and ensure accountability effectively," without which current deficits will continue to compound (Vanhontegem 2023). Yet, overcoming these challenges can be very difficult in the short term as poor institutional arrangements and coordination go hand in hand with a lack of governmental leadership, awareness, and commitment by officials; harmony related to the roles and responsibilities of sectoral actors; mechanisms to bring different stakeholders together; collective action, linking the different levels of government and service providers; and a lack of institutional sustainability (Agenda for Change 2023).

Institutional arrangements and coordination are weak when the institutions over policy, planning, service provision, and service authority either do not exist, need to have clearly defined mandates and roles, or have insufficient capacity and resilience. The absence of coordination mechanisms leads to the need for more harmony between WASH actors along with coordination across other sectors (Agenda for Change 2023). According to the ICRC and Norwegian Refugee Council (NRC) (2023), without addressing institutional shortfalls throughout the Middle East, governance systems are weakened which can result in institutions "los[ing] capacity, resources and organizational knowledge, get cut off from international efforts and be unable to include civil society and private-sector organizations in the management of common resources because security is prioritized over other issues, or because of a lack of trust" (Darbyshire et al. 2023). The foundation of long-term resilience and sustainability is built on correcting these shortfalls.

Resilience Building

Resilience building is the process of strengthening institutions and helping individuals, families, and communities prepare for shocks in a way that allows them to either bounce back to a sense of normalcy or to adapt and pivot in a way that still meets their needs. During our interview with Omar El Hattab, he noted that there is a need for all actors to not only focus on improved longer-term sustainability but also adopt resilience thinking. With climate change and water scarcity, like other shocks, there is a need for a resilient WASH sector that can absorb shocks and bounce back. El Hattab warned that "shocks can also be human-made, such as conflict instability, or from natural causes. However, climate change is surfacing as one of the largest threats. The [WASH] sector at large is not prepared, not resilient enough" (El Hattab 2023). In the Middle East, the smallest changes in precipitation or water mismanagement can have dire consequences on individuals, families, and communities and is particularly felt by marginalized groups. Successful climate change adaptation in the Middle East will require significant shifts in the priorities and thought processes of humanitarian, development, and government actors.

Resilience building is, therefore, a strategy that requires both an acknowledgment of the asset individuals, families, and communities, including marginalized groups, bring to resilience programing and also a recognition of climate change's potential to push people beyond their natural ability to cope, by removing them from their networks of support. Andy Bastable highlighted the importance of all regional actors asking the vital question, "How can your response, and the population you are targeting, be more resilient to climate change through all interventions and sectors?" (Bastable 2023). There is a dire need for more examples of best practice programing that understand the risk factors of climate change and how resilience thinking can lead to better, more sustainable climate adaptation in the Middle East.

Resilience building underscores the necessity of building a resilient sector capable of absorbing shocks, whether from climate change or other sources.

A transition considering climate resilience and sustainability as central objectives is essential for lasting impact. Success will require developing a new humanitarian lens capable of simultaneously seeing short-term, lifesaving aid and long-term development. Omar El Hattab noted in our interview that better tools able to adequately measure sector resilience in humanitarian contexts are essential steps to developing effective and sustainable resilience programing and will be invaluable as actors and their institutions in the Middle East renew efforts to assess climate vulnerabilities and make vital changes to increase resilience and adaptability (El Hattab 2023). Yet El Hattab warned that

> preparedness is not good enough. We are never prepared for all scenarios. How prepared can we possibly be? So the best way is to try to ensure that the sector is able to absorb the biggest part of the shock so that we only need to deal with the residual.
>
> (El Hattab 2023)

El Hattab's statement acknowledges the complexity inherent in the growing challenges of climate change. It highlights the increased need for greater integration across the HDP nexus as actors seek to address the growing crisis.

The WASH sector in the Middle East needs to assess carefully how resilient it is to climate change. Growing evidence shows that high levels of WASH services do not necessarily correlate with high resilience to shocks (Batchelor, Smits, and James 2011; UNICEF and GWP 2022; World Bank, ICRC, and UNICEF 2021). WASH systems must be specifically designed with climate resilience in mind, and as Batchelor, Smits, and James point out, "some widely-used WASH technologies may not have the potential to improve resilience under some scenarios" (Batchelor, Smits, and James 2011, 26). Careful consideration of coming risks must be matched with new and future innovations and technology to create sustainable services. Access to safe water is of limited benefit if that water supply is irregular or easily shut off by unreliable power grids or upstream water usage. Yet the benefits of well-thought-through resilience programing can be significant, as a report by UNICEF and the GWP (2022) highlighted, noting that "resilient WASH services have positive economic impacts on household incomes in the short, medium and long term, help reduce the risk of migration and conflict, and create more resilient environments against the spread of global health pandemics" (UNICEF and GWP 2022, 28). Climate reliance is, therefore, built upon anticipatory programing that focuses on providing services and decreasing future vulnerabilities.

Institutional Regulations, Access, and Inclusion

Addressing WASH needs in the Middle East is also significantly hindered by poor or non-existent regulations and policies. This is particularly true when considering current regulations and policies relating to climate change and coordination across the HDP nexus. During our interview with Khaldon Khashman, he stated,

> Institutional regulations [either non-existent or those currently being implemented] are among the greatest threats to addressing WASH needs in the Middle East, taking into account the considerations around whether or not they do or do not meet the needs of citizens and their workers. I think we need to modify regulations and registrations in order to improve the basic standards within the sector.
>
> (Khashman 2023)

The lack of robust regulations addressing climate change in the Middle East presents a multifaceted challenge with profound implications for humanitarian and development practitioners operating in the region. This deficiency in regulatory frameworks creates a vulnerability that permeates various aspects of environmental management, resilience, sustainability, and adaptation efforts. This was a key theme found across both our survey respondents and interviewees, where respondents highlighted the lack of regulations in the region as a major threat to the WASH sector. According to a survey respondent who works with UNHCR in the Middle East, "There is an absence of regulations for water resources management and environment protection at a time when the climate vulnerability is on the rise." Another survey respondent working with the French Water Partnership, speaking specifically about sanitation, noted, "There are few efforts provided by institutions in preserving water and its contamination." Furthermore, overcoming specific environmental challenges that come about from climate change, as argued by a respondent working with UNICEF and the GWC, "will require policy change[s], and regulatory reform[s], which is exceptionally challenging when working in fragile states." The absence of specific regulations tailored to climate change in the Middle East hinders the effective management of water resources and environmental protection while weakening their long-term resilience to future shocks (Waha et al. 2017).

For instance, southern Iraq, once water-rich, now faces severe droughts and significant drops in annual rainfall due to upstream dam constructions in Turkey and Iran. This has led to a drastic decline in water levels in Basra (in the south of Iraq), turning it from the "Venice of the Middle East" into a region with undrinkable water and widespread hospitalizations due to salt contamination (El-Geressi 2020). Similarly, climate change in Alexandria, Egypt, is causing sea levels to rise, threatening the city's infrastructure, agricultural lands, and cultural sites as it sinks from the north. These examples from Iraq and Egypt underscore the critical need for climate change regulations to effectively manage environmental and water crises in the Middle East (El-Geressi 2020; Kamal et al. 2021). Climate change introduces new dynamics, such as altered precipitation patterns, rising temperatures, and increased frequency of extreme weather events (Alaaldin 2022; Baxter et al. 2022). Without targeted regulations, these changing conditions make water resources management increasingly challenging, sometimes leading to conflict and regional struggles to address issues like water scarcity, quality preservation, and ecosystem protection.

Our survey respondents also hinted at how a lack of regulation on climate change specifically impacts the humanitarian and development nexus. For humanitarian practitioners, this regulatory gap complicates disaster response and preparedness efforts. The lack of clear guidelines and standards from local and national decision-makers regarding climate change leaves organizations without a comprehensive framework to address the evolving needs of communities affected by climate-related disasters. In extreme weather events or prolonged droughts, the absence of regulations leads to delays in response, inadequate resource allocation, and difficulties in coordinating relief efforts. Moreover, from a development perspective, the deficiency in climate change regulations poses obstacles to long-term planning and infrastructure projects. Without a regulatory framework that considers climate resilience, development initiatives often fail to account for the escalating challenges of climate change. These failures, often due to inadequate resilience-building efforts during program design and implementation, result in vital investments in infrastructure that become quickly outdated or insufficient due to unexpected climate-related disruptions.

The lack of regulation around climate change also affects the accountability and commitment of governmental bodies. Humanitarian and development practitioners often collaborate with local governments to implement projects and initiatives. When there are no clear regulations guiding climate action, governments may lack the necessary incentives to prioritize climate-related issues, hindering the integration of climate considerations into broader humanitarian and development agendas. In many cases, humanitarian and development actors bypass institutional bodies altogether, removing local and national governments from vital data necessary for effective resource management (Tillett et al. 2020).

As a result of this inaction, humanitarian and development agencies sometimes take on roles as climate change champions in place of government inaction. For example, the Middle East is a fossil fuel-rich region. However, investments in other non-renewable energy, for example, are sometimes prioritized and spearheaded not by the governments themselves but by the international organizations working there. This presents a long-term challenge of dependency in the low-income nations of the Middle East as these communities also heavily rely on humanitarian funding to address the challenges posed by climate-related risks. This questions the responsibilities of ownership, commitment, and accountability on the government's side in taking action and imposing the regulations necessary to combat climate change in place of other actors (Duenwald et al. 2022).

It should be noted that the provision of essential WASH services in the Middle East faces institutional challenges and access constraints rooted in security concerns that are exacerbated by conflict, political instability, and operational limitations. These challenges significantly impede the ability of implementing partners to deliver timely and effective WASH solutions, resulting in delayed responses, extended project timelines, and compromised outcomes. This was brought up during our interview with Denis Heidebroek, who noted that

In places such as Syria and Yemen, there are many locations that are home to vulnerable populations in need that are unfortunately off limits to humanitarian agencies to intervention because of security constraints. This poses a major challenge to actors and partners on the ground, and their ability to implement within a given time frame. In some instances, it is impossible to mobilize money within the given timeframe, as agreed upon by implementing partners and donors, resulting in project extensions. Most importantly, project extensions result in the most affected populations not being reached when they are at their most vulnerable and in need of assistance.

(Heidebroek 2023)

These constraints are genuine and significantly complicate the already challenging issues around climate change mitigation and adaptation, humanitarian response, and broader resilience-building efforts. Yet these challenges are rooted in political and institutional failures, which require political and institutional solutions.

Bureaucratic impediments, including complex visa and permit processes, delay or prevent timely aid delivery, further complicating humanitarian access. Additionally, violence and insecurity stemming from political instability, including direct attacks on aid workers and vital infrastructure, pose a severe risk, making it difficult to provide aid safely. These challenges significantly impede the work of humanitarian and development agencies, as they restrict movement, limit access to critical infrastructure like hospitals and schools, and create financial and logistical burdens. This not only hampers the delivery of immediate aid but also undermines short- and long-term development efforts and the agencies' ability to operate effectively and reach those in dire need (Kurtzer 2019).

Adding to these obstacles are various factors that contribute to the situation's complexity, most notably, inflexible funding mechanisms that restrict the adaptability of implementing partners, hindering their capacity to respond dynamically to evolving needs. Furthermore, the lack of proactive risk assessments leaves projects susceptible to unforeseen challenges. Reactive approaches undermine the development of robust strategies to address potential obstacles in delivering resilient and sustainable WASH services. Compounding these issues is the need for more information sharing among actors on the ground and donors. This communication gap hampers coordinated efforts, impeding the overall effectiveness of WASH initiatives in the region. A significant gap exists in building meaningful and lasting relationships with local authorities and communities, undermining the collaborative nature of WASH projects. Establishing strong partnerships is crucial for navigating local contexts and ensuring the sustained success of climate-related mitigation and adaptation efforts. Lastly, failing to adequately share information with local communities regarding the importance of WASH interventions further weakens local institutional strength. It impedes the overall impact of both climate change and WASH initiatives. Enhancing community awareness and involvement emerges as a pivotal factor for the success of WASH programs in the Middle East and requires local and national institutional leadership to succeed. Addressing these multifaceted challenges requires a comprehensive approach considering the immediate access constraints and the broader institutional shortcomings across the Middle East (United States Agency for International Development 2022).

Beyond the physical access constraints faced by HDP actors and donors, there are increasing issues around equal rights and inclusion as efforts to ensure that marginalized and vulnerable populations have access to the services and resources they need and deserve. The impact of climate change is felt the greatest by marginalized groups who lack the resources and skills necessary to adapt to current climate-related shocks adequately, let alone future anticipated pressures. In our interview with Denis Vanhontegem, he stated that

> One significant institutional challenge lies in establishing and advocating for equitable access to essential services for the most vulnerable populations and understanding how this equitable access contributes to maintaining peace, especially in the context of climate change and the transition toward the humanitarian-development nexus.
>
> (Vanhontegem 2023)

Finding and harnessing the political and institutional will to effectively address these issues remains challenging. The climate change problem is complicated by institutional weaknesses and conflicting local and national priorities competing for inadequate resources.

Yet, the challenge of equal rights and inclusion regarding access to resilient and sustainable WASH services is further underscored by the disproportionate impact on vulnerable populations, particularly women and girls. The unequal distribution of water resources, coupled with issues such as displacement, conflict, weak governance, and political unrest, amplifies challenges for marginalized and impoverished members of society. As urbanization accelerates across the Middle East, water scarcity exacerbates existing disparities, disproportionately affecting vulnerable groups, including the elderly, women, children, ethnic groups, and people living with disabilities.

The implications are especially severe for women and girls, given their crucial roles in livelihoods, food security, and water-related responsibilities. According to a recent UNICEF and World Health Organization (WHO) report, women and girls in 7 out of 10 households without water supplies on the premises are primarily responsible for water collection, highlighting the gender disparity in WASH access (UNICEF and WHO 2023). It's important to note that gender-based violence often increases in contexts where women and girls are vulnerable due to a lack of access to essential services, including WASH. Inadequate WASH facilities can exacerbate the risks of violence, especially for women and girls, by forcing them to travel to remote or unsafe locations to access water or sanitation, increasing their exposure to potential harm. As a result, women will be put at further risk as climate change forces them to travel to even more remote and unsafe water sources to meet their WASH needs and those of their households. The intersectionality of gender-based violence, menstrual hygiene needs, and specific challenges faced by women and girls in accessing clean water and sanitation further emphasizes the urgency of inclusive and gender-sensitive water management approaches (Akacha 2023).

Similarly, the UNICEF and WHO report highlights stark inequalities in WASH access, with a particular burden on women and girls, calling for integrated gender considerations in WASH programs and policies (UNICEF and WHO 2023). There is a need for a continued effort to advocate and promote embedding cross-cutting issues, specifically around gender, gender-based violence, disability inclusion, accountability to affected peoples, prevention of sexual exploitation and abuse, age, environment, and climate change, in targeting the people most affected by and vulnerable to crises, in response planning and monitoring (Gray et al. 2022). In a region facing social disparities intensified by climate-related stresses, prioritizing equal rights and embracing gender-sensitive, inclusive strategies in water resource management and WASH services are crucial for successful adaptation and resilience building. Ensuring equitable distribution of services and resources, especially during crises, is fundamental to these efforts. These considerations are critical to effective climate-related resilience-building efforts and require greater local and national institutional leadership and collaboration.

Leadership and Collaboration

At its core, the key problem currently faced in the Middle East's response to climate change is ineffective leadership, due mainly to weak local and national institutional regulations and frameworks that lead to confusion in how to address the challenges of climate change more cohesively. Farah Al-Basha argued further in our interview that,

> It is clear that several countries are having droughts in the region, and there isn't enough being done to plan and prepare for this change. The question for the WASH sector is, where do we start? Is it establishing early warning systems or understanding groundwater levels? Do we start with establishing preparedness plans? What is the level of resistance from the government in addressing climate change? Is it the role of the government or the UN to create preparedness plans? If we do not ask and answer these questions in advance and simply jump to respond when the disasters occur, we cannot do much. It is too late.
>
> (Al-Basha 2023)

In essence, there is no clear plan in place for all actors to follow, and local and national governments, by and large, currently lack the leadership, resources, capabilities, and foresight to adequately mitigate and adapt to climate change's current and future challenges. Institutions need a clear vision of the roles and responsibilities of actors working within the water management and WASH sectors. They also do not have an outline for what questions to ask and what issues to address first in preparation for climate change disasters or repercussions. Farah Al-Basha noted during our interview that "there is a need for clear position papers and stakeholder mapping to see where we can start and sustain initial engagement based on what is needed and by whom" (Al-Basha 2023). Al-basha specifically suggests that addressing WASH needs in the face of climate change across the humanitarian-development nexus requires well-defined documentation outlining the positions, objectives, and strategies of relevant humanitarian and development actors by local and national governments.

Additionally, conducting stakeholder mapping is necessary to identify key players and their roles, helping to determine where and how to begin and maintain effective engagement based on each stakeholder's specific needs and responsibilities. This emphasizes the importance of institutional planning and coordination to ensure a coherent and sustainable approach to tackling climate-related WASH challenges. Climate change is a conflict and risk multiplier, yet there is a fundamental disconnect between acknowledging the breadth of the problem and connecting challenges to concrete, actionable steps forward. Instead, because of the complexity of the problem, climate change does not make its way into the national agendas throughout the region. As a multiplier, it can inflict immense hardship on a region already beset by conflict, social upheaval, and economic woes (Baxter et al. 2022). In the meantime, the growing gap in climate change planning and positions across the Middle East requires societal and governmental culture shifts that permit critical reevaluations and public sector reform (Waha et al. 2017). Adapting and adopting good governance techniques that promote and facilitate innovations to mitigate climate-related issues, supported by research, the private sector, and raising public awareness, is critically important to building current and future WASH services (Alaaldin 2022).

Developing institutions that will correct current weaknesses and confront future challenges requires a road map to stronger local and national leadership, which will result in stronger, more impactful collaboration across the HDP nexus. To date, one of the strongest frameworks for confronting water resource and WASH service challenges has been the development of the JOF, which outlines guidance around identifying institutional weaknesses and designing and framing coordination efforts that build resilient and sustainable services moving forward. The JOF is uniquely suited to act as a unifying starting point for institutional reform and collaboration across the HDP nexus. Although not a new initiative or a silver bullet solution, the JOF simplifies the complexity of reform so that coordination becomes feasible even in areas of fragility where local and national institutions are weak or seemingly nonexistent.

Climate change poses a unique challenge as it is often the cause of conflict and instability and the source preventing peace (Wong et al. 2020). Yet, as local and national decision-makers and institutions in fragile states increase resilience and adaptation efforts, the JOF is ideally suited to aiding actors already facing the unique challenges of climate change. Timothy Grieve went on to state the

> Experience of implementing the nexus approach suggests that it is particularly relevant in contexts prone to protracted (e.g. long-term conflict, water insecurity, etc.) and recurrent (e.g. floods, drought, etc.) crises. In such contexts, the HDP pillars collaborate to provide the required stability to implement long-term programmes. Given the role of climate change as a driver of conflict and protracted/recurrent crises, the JOF is also highly relevant in climate-vulnerable contexts.
>
> (Grieve 2023a, 36; Grieve, Panzerbieter, and Rück 2023, 3)

Achieving long-term climate change outcomes will require political will and the coordinated efforts of governments with their humanitarian and development partners. This sentiment was reiterated during the 6th Annual Arab Water Week in March 2023, a yearly event organized and run by the Arab Countries Water Utilities Association (ACWUA) and a key regional conference where leaders and organizations in WASH from across the Arab region gather to discuss, collaborate, and advance WASH outcomes. ACWUA's final report from the event emphasized inspiring Middle Eastern governmental and institutional leadership to address climate change, focusing on the unique challenges around water scarcity in the region (ACWUA 2023). Yet, this conclusion that greater regional leadership is needed at both the local and national levels has been articulated consistently as the vital step to unifying climate response strategies (Agenda for Change 2023; Sanitation and Water for All 2020a, 2020b). The WASH Agenda for Change, a collaboration of organizations working to strengthen WASH systems globally, similarly highlights that the building block of institutional arrangements is strong when

> the institutions related to sector policy and planning, service provision, and service authority exist, have clearly defined mandates and roles, and have sufficient capacity. Coordination mechanisms are in place, leading to harmonized action within the WASH sector and with related sectors.
>
> (Agenda for Change 2023)

It is clear that addressing political instability and strengthening local and national institutions are critical to any lasting climate response in the region. Without the necessary infrastructure and stability to ensure access to clean water and adequate sanitation for all people by 2030, attaining SDG 6 may become unattainable (UN DESA 2017).

With institutional leadership comes the need for coordination across the HDP nexus and the ability to bring key stakeholders, particularly marginalized and underrepresented groups, to the table of climate adaptation decision-making, thus allowing for greater harmony in actions and outcomes (Agenda for Change 2023; Sanitation and Water for All 2020a, 2020b). Utilizing the JOF is an easy first step for even the weakest local or national institutions to build leadership and collaboration capacity. The JOF is a pivotal tool in strengthening institutions within the WASH sector, particularly in regions grappling with the complexities of climate change, conflict, and fragile institutional structures. The JOF, by its very design, helps foster a collaborative approach across the HDP pillars. It emphasizes the need to integrate resilience, conflict sensitivity, and peacebuilding into WASH programs, thereby enhancing the capacity of institutions to handle crises effectively and sustainably. Moreover, the framework advocates for strong institutional systems and community involvement, which is essential for reducing long-term humanitarian needs and achieving sustainable development in WASH services. This holistic and inclusive approach ensures that WASH interventions are effective, resilient, and adaptable to the changing needs of vulnerable communities (Grieve, Panzerbieter, and Rück 2023).

Water scarcity is a critical topic across the Middle East. Yet, more institutional and legislative discussions and action are required to understand better and protect water's true value, including the social, environmental, and financial costs of extraction and delivery of WASH services. Integrative government approaches will require stronger linkages between multisectoral decision makers as agricultural, educational, healthcare, environmental, and social actors seek to sustainably utilize and protect scarce water resources. Bringing these actors together productively will require coordination, cooperation, and eyes focused on innovation. Lastly, compromise will be essential to solving cross-border water disputes. Focusing on people's needs and rights will have to prevail over national resources security in diplomatic discussions and agreements to successfully provide sustainable and resilient, long-term water resource management and safe and reliable WASH services to all people in the Middle East (RBAS 2013).

In other words, there is a dire necessity for robust legal frameworks and policy shifts to ensure sustainable water resource management. The region's vulnerability to climate change necessitates adaptable regulations considering changing water patterns and resource availability. Furthermore, there is a need to align policies with the evolving nexus approach that is essential for effectively integrating humanitarian and developmental efforts. Overcoming these institutional barriers requires collaborative efforts to reform regulatory frameworks, enhance cross-sectoral coordination, and bolster legal mechanisms for equitable and resilient WASH services (Fantini 2019). This coordination will require more effective tools, platforms, and coordination frameworks as humanitarian, development, and national actors seek to combine resources, innovation, and objectives to meet local needs better (UNICEF 2022). Yet, as Neira et al. (2023) noted, this coordination must be regional in scope to succeed, requiring countries, along with their humanitarian and development partners, to share data around successful, low-cost technologies and their impacts on decarbonization, water access, environmental and healthcare outcomes (Neira et al. 2023).

We recognize that various political realities and instability due to conflict remain a significant barrier to meaningful implementation of climate change adaptation. This can result in delayed responses, extended project timelines, and compromised outcomes. Yet here, we agree with the United States Agency for International Development's (USAID) Global Water Strategy 2022–2027, which suggests that adopting flexible funding mechanisms, establishing proactive risk assessments, information sharing, coordination, and joint problem-solving is essential for reaching desired outcomes. This will require prioritizing building strong relationships with local communities and leaders, leveraging their insights and influence to enhance access and acceptance of WASH interventions, which will greatly increase the long-term ability of humanitarian and development organizations to reach marginalized groups with the most need (United States Agency for International Development 2022). Overcoming the challenges posed by access constraints requires innovative approaches from donors, implementing partners, and governments that ensure safe and unhindered access to humanitarian and development interventions, enabling them to deliver urgently needed WASH assistance to vulnerable populations in a rapidly changing context.

Governance issues within water authorities become critical in ensuring accountability and effective management. Furthermore, the lack of preparedness of many governments to respond proactively to climate change and natural or artificial disasters creates a precarious scenario that can readily devolve into humanitarian crises. The fundamental issue of institutional arrangements and coordination becomes evident as a cornerstone for the successful functioning of the WASH sector. The absence of clearly defined mandates, coordination mechanisms, and harmonized action within the sector can result in uncoordinated efforts, undermining the sector's capacity to meet WASH needs effectively. Climate change remains a multi-institutional and multi-sectoral challenge that requires strong leadership, coordination, and the collective action of public ministries, institutions, and actors working towards short-term and long-term environmental and climate resilience. Weak institutions, inadequate coordination, and a lack of responsive governance systems hamper the region's capacity to address WASH requirements efficiently. If these institutional shortcomings are not effectively addressed, they could perpetuate a cycle of inadequate resources and uncoordinated actions and effectively respond to the multifaceted institutional challenges faced by the Middle East.

4 Transitional WASH across the Nexus

Climate change is one of the largest protracted crises in the Middle East and poses a significant challenge to humanitarian and development actors considering how and when to shift from short-term, lifesaving aid to long-term developmental initiatives. The gap between when humanitarian action ends and when development begins needs to be better defined. This gray area can lead to confusion and infighting yet, critically, can also catalyze further collaboration and actors seeking to better meet the needs of people living in areas of fragility and crisis. This transition is further complicated by international norms that often place whole regions or countries under the banner of humanitarian context. Yet, contexts on the ground can vary dramatically across crisis areas, with some regions holding onto greater stability better than others. This can lead to fractures where humanitarian assistance is needed in areas where a neighboring location may be capable of greater developmental programing. Developing stronger coordination between humanitarian and development actors is essential to addressing the complex immediate and long-term needs caused by climate-related pressures in the Middle East. As argued in the last chapter, greater local, national, and regional leadership is needed to allow for effective transitions in the face of significant regional differences, needs, and context.

Much of our discussion in this chapter requires a lens much larger than just transitioning WASH interventions from humanitarian to development actors. Yet WASH programing plays a unique and vital role in addressing the impacts of climate change in the Middle East due to its deep-rooted and multisectoral connections to other sectors such as health, environment, agriculture, livelihoods and food security, education, shelter, and child protection. Additionally, WASH systems are critically fragile to climate-related shocks, and, as discussed in previous chapters, poorly maintained WASH services are strongly linked to cycles of protracted conflict, forced migration, and socioeconomic instability. These cycles particularly impact marginalized groups who need more resilience capabilities and resources to adapt to the unique pressures posed by climate change and require assistance that many local and national systems need to be prepared to render. As such, WASH services provide a

DOI: 10.4324/9781003436706-4

unique entry point for climate action in the Middle East as local, national, and regional actors seek to mitigate the compounding impact of water scarcity and population growth in the region and develop adaptive interventions that build resilient and sustainable outcomes for increasing numbers of at-risk and marginalized groups. Our research, interviews, and surveys in the region indicated the need for larger system changes and greater commitment to integrated collaboration and programing across the HDP nexus.

Across the Middle East, crises are only growing in length and scale. The current reality is that very few conflicts or situations are short, and many refugee and internally displaced persons (IDP) camps, initially intended as short-term solutions, are still in existence decades after their establishment. This raises the question of humanitarian actors' role in delivering longer-term interventions from day one. These conclusions were raised in several of our interviews with practitioners and thinkers across the region. In an interview with Andy Bastable, he stated, "In these more frequent instances of protracted crises, there is a need to focus on transitional and longer-term WASH. Looking at the longer-term objectives from the beginning is a real need. Not enough is being done in this regard" (Bastable 2023). Yet, even with pressure to shift humanitarian and development thinking to a more integrated mindset with the New Way of Working and the Grand Bargain (A4EP 2021; Center on International Cooperation 2019; IASC 2021; ICVA 2017; Nakamitsu et al. 2017; OCHA 2017; United Nations 2017), humanitarian actors continue to miss opportunities to think long term and persist with traditional short term, and increasingly short-sighted programing, often with long-term consequences. During our discussion with Andy Bastable, he explained that,

> For example, in the new refugee camps in NW Syria, humanitarian actors have only short-term thinking in addressing immediate WASH needs. Are actors also thinking about the aquifer levels? Not all emergency responses are thinking about the aquifers, anticipating future flooding, or saving water in anticipation of future needs. Not enough humanitarian actors are thinking about these longer-term concepts. There is a dire need to avoid the issues of humanitarian heroes who do quick fixes and leave.
>
> (Bastable 2023)

The transition across the HDP nexus is growing in length and now requires redesigning how HDP actors achieve long-term resilience and sustainability in program outcomes. Heroes with good intentions need to be replaced with intentionality and long-term investment.

Although much of this chapter will look at the specific needs of transitioning and integrating across the humanitarian-development nexus, it is only one part of a large, ongoing discussion around the HDP nexus. The HDP nexus is an approach that seeks to integrate HDP efforts to address crises and conflicts more comprehensively, enhancing the effectiveness, efficiency, and sustainability of interventions in complex and protracted crises. This involves capacity building, community engagement, and partnerships with local authorities to maintain and enhance vital infrastructure that fosters stability and peace. The HDP nexus encourages a holistic approach, recognizing that the interplay of HDP initiatives is essential for comprehensive and enduring solutions. Land and Hauck (2022) noted that

> The value-addition [of the HDP nexus) is both in terms of the "lens" it offers to enable a more complete and shared understanding of context across services, as well as in terms of the modus operandi it promotes towards setting common objectives, working in a coordinated or more joined-up manner, in promoting dialogue and information sharing across services. In doing so it also facilitates a better appreciation of how each service operates including the policy and institutional frameworks each has to work within.
>
> (Land and Hauck 2022)

Climate change acts as a compounding factor, stressing HDP actors, which necessitates collaborative efforts by humanitarian and development WASH actors to address the complex impact of climate change comprehensively.

The operationalization of the HDP Nexus faces challenges in defining its purpose and objectives without comprising institutional mandates, hindering the effective integration of strategies and actions. Clear leadership, joint planning, and programatic coordination are crucial at the national, regional, and country levels to overcome structural obstacles and promote a more inclusive and result-oriented approach. Operational challenges involve transaction costs, potential uncoordinated initiatives from various development partners, risks of increased bureaucracy, limited substantive capacity, and uncertainties about funding instruments. Beyond commitments, more concrete action is needed to push the nexus agenda forward, including improving inclusivity among affected populations and supporting multi-year funding in protracted crises. These measures are essential for bridging the humanitarian-development divide and ensuring lasting peace and more effective and sustainable responses to the complex challenges faced in the Middle East (ICVA 2022; Land and Hauck 2022).

Integration discussions take many forms yet highlight the multisectoral nature of crisis and disaster response. Another critical example of cross-sectoral coordination between the humanitarian and development sectors includes the Water, Energy, and Food Nexus (WEF) nexus, which is a concept that recognizes the interconnectedness and interdependence of WEF systems. The WEF nexus is critically important to climate change adaptation and mitigation efforts, particularly in the Middle East. It emphasizes the need for greater integration across the WEF nexus, utilizing resilience and sustainable approaches to managing these vital resources as changes in one sector can significantly impact the others. For example, water is essential for agriculture (food production) and energy generation, and energy is required for water pumping, treatment, and irrigation, creating a complex web of dependencies.

Understanding the WEF nexus becomes crucial in the context of the humanitarian and development nexus. In humanitarian situations, where crises often disrupt these essential systems, a holistic approach to addressing the immediate needs of affected populations requires consideration of these interconnections. In a development context, long-term planning must incorporate strategies that balance the use and management of WEF resources to ensure sustainability and resilience. Recognizing the nexus helps bridge the gap between emergency response and sustainable development, fostering integrated solutions and considering the broader implications on communities' well-being and ability to withstand future shocks.

In alignment with a focus on transitioning across the humanitarian-development nexus, other studies focus more on aspects around the WEF nexus in a humanitarian context. One such study by Srivastava et al. (2022), which focused on a case from the Rohingya refugee camps in Bangladesh with an emphasis on how household-level access to WASH services, also considered the use, access, and availability of energy and food in addition to their effects on host-refugee interactions. The results from the study reveal that there is an implicit and explicit link between WASH and WEF. Even a small intervention in any WEF area positively affects the other resources, especially in enhancing resource access and use. Furthermore, there are often complex, indirect interactions between WASH and the other components of the WEF nexus. Highlighting this point, Srivastava et al. (2022) argued further that WASH "blends the social dimensions (access, safety, consumption, and use) with the WEF resource dimensions (availability and resource sustainability), including a further emphasis on sanitation as a key, but often ignored, element of the WEF nexus" (Srivastava et al. 2022, 4). Critically, in humanitarian contexts, bottom-up perspectives on these interlinkages with active participation from host and refugee households are required to understand the implicit and explicit connections across WASH and the WEF nexus (Srivastava et al. 2022).

Transitional discussions around the HDP and WEF nexus are critical in the Middle East, where climate change's realities significantly impact the duration and spread of crises as conflict around water, energy, and food continues to increase. This is particularly true regarding the scarcity of water and energy resources, exacerbated by conflict and directly affecting humanitarian and development interventions. For example, the need for energy to power water pumping and treatment systems and the demand for water-intensive agriculture linked to food production create a complex dynamic. Both humanitarian and development actors require a deeper understanding of food and water security repercussions. Together, they should focus on their work's impacts on land, water, and energy resources, ensuring natural resources and ecosystems can sustain the transition from humanitarian assistance to development programing. These interdependencies highlight the importance of ongoing discussions around the HDP and WEF nexus in developing transitional initiatives to ensure sustainable and resilient solutions to climate change's current and future challenges.

Yet the humanitarian-development divide has long been a contentious debate in both academia and among international agencies and governments. Despite the recent surge in the frequency, duration, and severity of humanitarian crises, humanitarian and development disciplines and communities of practice, despite calls for reform, have continued to default to siloed responses, which perpetuate ineffective practices in crisis response. (Blind 2019; Snel and Sorensen 2021, 2023a; Sorensen and Snel 2022). Operating within silos emphasizes separation, neglecting where each discipline has areas of connection and overlooking areas where their interventions overlap. The few actors who adopt their own perspectives on the HDP nexus often rely on generalized approaches in the absence of defined and agreed-upon activities and outcomes, whereby the context-specific nature of the nexus is often overlooked (UNRISD 2020).

Bridging the humanitarian and development nexus is a phase in which a community transitions towards a response paradigm that combines humanitarian response with long-term services. Historically, most studies have focused on either humanitarian or development WASH services, which is a point that was reinforced in our interviews and surveys. Only in the last few years has a greater focus been placed on transitional WASH programing. One such study on the implications of the transition between the emergency and protracted stages of displacement focuses on its impact on environmental health conditions. Cooper et al. noted that,

> In 2019, 30,000 people were forced to leave their homes due to conflict, persecution, and natural disaster each day. Eighty-five percent of refugees live in developing countries, and they often face underfunded and inadequate environmental health services. Many displaced persons live in camps and other temporary settlements long after the displacement event occurs. However, there is little evidence on environmental health conditions in the transitional phase – defined by the UNHCR as six months to two years after displacement. Based on their findings, water supply was the most frequently discussed environmental health topic. Overcrowding was the most common risk factor reported, *Vibrio cholerae* was the most common pathogen reported, and diarrhoea was the most commonly reported health outcome.
>
> (Cooper et al. 2021, 1)

Cooper et al. emphasized the need for increased collaboration among government and non-governmental organizations to improve resource provision during this critical phase of displacement and address significant knowledge gaps (Cooper et al. 2021). This article provides one of the first comprehensive reports on environmental health conditions in the transitional stage, which is essential in ensuring the success of the humanitarian and development nexus transition.

Another study conducted by Bakchan et al. (2021) delved into the challenges of the humanitarian and development nexus, particularly focusing on municipalities in Lebanon across physical, social, financial, and institutional dimensions. The research addressed the difficulties water and wastewater utilities face in providing services to displaced populations after crises. The study advocated for resilient water and wastewater infrastructure systems, proposing an integrated approach that acknowledges the humanitarian and development nexus. Through interviews with municipalities in Lebanon, the research identified contextual challenges and explored interactions to inform effective solutions. The study's findings emphasize the importance of introducing policy measures, such as utility pricing and shared development priorities, to facilitate the transition across the nexus transition and achieve resilient systems, contributing to broader discussions on sustainable WASH development (Bakchan, Hacker, and Faust 2021).

Finally, a 2017 study by Martina Rama focused on linking relief, rehabilitation, and development (LRRD) together, which aims to improve integration and coordination between humanitarian and development actors in transitional contexts. LRRD is an approach that emphasizes the interconnectedness and seamless transition between humanitarian relief, rehabilitation, and long-term development efforts in crisis or conflict-affected areas. Instead of viewing these phases as separate and distinct, LRRD aims to create a continuum where emergency relief activities address immediate needs and contribute to the affected communities' long-term recovery and development. This approach recognizes that a well-coordinated and integrated response can help break the cycle of crises and build resilience, fostering sustainable development outcomes (Rama 2017). In the study, Rama emphasizes the need for aid practitioners to invest significant time and resources in initial need assessments, balance ambition and pragmatism in choosing WASH infrastructure, and allow flexibility in project methodologies to adapt to evolving circumstances. These recommendations highlight the practical aspects of transitioning from emergency responses to more sustainable development interventions, addressing the challenges of unstable environments (Rama 2017). Overall, the report adds to the ongoing discourse around bridging the gap between humanitarian and development efforts, offering lessons learned and practical recommendations to enhance the effectiveness of interventions in fragile states and protracted crises.

Although much of this book focuses on transitional WASH from humanitarian to development, the same issues in ensuring a seamless transition can be found in instances where there is a transition in reverse, as a development context slips into a humanitarian context, such as in Yemen in 2015. A study by Mena and Hilhorst (2022) investigates the transition from development to humanitarian aid in Yemen amid escalating conflict, focusing on disaster-related actions, particularly addressing water-related hazards like drought and water scarcity. Before the crisis, water programs were part of general WASH and development initiatives, often developed in partnership with the Yemeni government and designed as long-term solutions. However, since the crisis, water-related issues shifted to being addressed within the general humanitarian response, primarily focusing on water delivery by tanker trucks (Mena and Hilhorst 2022). The study reveals a significant shift in actors and agencies, with many international development organizations scaling down operations, leading to the rise of Yemeni NGOs handling WASH interventions in their absence.

The findings highlight the challenges during the crisis, with local NGOs leading most responses but lacking preparedness for large-scale humanitarian efforts. Development projects modified during the humanitarian and conflict period were typically initiated by local actors with prior development experience. The study underscores the need for better integration between pre-crisis development projects and humanitarian aid interventions, emphasizing the importance of continuing development work during crises. Challenges in the transition include inflexible programing, limited space for national actors to set agendas, and changing donor policies responding to conflict escalation. The research calls for a more flexible approach in humanitarian aid, incorporating development-related and disaster risk reduction actions during emergencies, particularly in the WASH sector, to bridge the artificial divide between humanitarian and developmental aid for more effective crisis management (Mena and Hilhorst 2022).

Developing a Nexus Mentality

These recent studies connect well with our survey and interview findings and call for a stronger commitment from HDP actors to think, design, and implement programing with long-term outcomes in mind. This is particularly true when addressing the issues of water scarcity, WASH services, and climate change, which are heavily interlinked. During our interview with Daniel Karine, he stated, "A key question is how climate change will impact the population that humanitarians are targeting in the long run, and how to better prevent such changes and be better prepared. Time is the parameter here" (Deniel 2023). Deniel here is suggesting that even for those organizations who are not active in transitional nexus discussions, it is vital for them to adopt longer-term thinking and ask themselves how they are contributing to the long-term resilience of the WASH sectors as a whole and the ability for local and national decision-makers to address the current and future challenges posed by climate change. Humanitarian actors cannot focus only on day-to-day interventions but must start early to lay the building blocks for future interventions. This point was driven home during our interview with Anna Rubert when she stated, "We cannot wait for the establishment of strong institutions to take on this role, notably in conflict. We cannot afford to wait, as crises become increasingly more protracted throughout the region" (Rubert 2023).

The lack of longer-term, transitional nexus thinking found further support from our survey respondents, including a member of the French Water Partnership, stating "The goal of the nexus should be to bridge humanitarian needs towards sustainable service provision in the long-run, taking into account climate change," suggesting the vital addition of long-term resilience and sustainability as a critical addition to the traditional humanitarian mandate to save lives. Without this critical, long-term consideration by humanitarian actors, a survey respondent from ACF cautioned that humanitarian organizations could inadvertently foster worse outcomes in their initial efforts. This results in interventions that cannot be sustained and in governments lacking the ability to assume leadership when these organizations withdraw. Moreover, even when donors and humanitarian entities aim to adopt a more extended-term perspective, they may not always possess the necessary resources and expertise, potentially leading to additional difficulties during transitional phases. Ultimately, humanitarian thinking and systems need to be redesigned to incentivize greater flexibility for humanitarian actors to both save lives and build resilience as a survey respondent from the ICRC argued stating that "The humanitarian-development nexus should allow for actors to design models that address, on the one hand, urgent needs, but at the same time, address the needs of essential services longer-term."

Yet bridging the humanitarian-development nexus doesn't just hit resistance from humanitarian actors' overreliance on short-term programing. Many development actors are overly risk-averse and resistant to entering fragile contexts too early. Another survey respondent from the London School of Hygiene & Tropical Medicine highlighted the need for more cross-collaboration between humanitarian actors and those leading long-term development programing. He noted that to address the shortfalls of improved nexus thinking, the nexus agenda should include more engagement with humanitarian actors to effectively and efficiently transition. This was also reinforced in our interview with Denis Heidhebroek, who highlighted the need for more cross-collaboration between the two sectors, stating,

> This will allow [humanitarian actors] to better prepare to know the grounds for development and developmental response and how it will take place and what they are doing. Engagement is very important, and adopting agreed-upon and shared tools, whether it be implementation plans, monitoring tools, and other such frameworks and templates that take into account the environment, sustainability and the nexus space altogether. This is something that some implementing partners do, in fact, subscribe to, but not all of them and their donors.

> (Heidebroek 2023)

Finding every opportunity to bring humanitarian and development actors in the same room simultaneously will allow for better, more streamlined program design, implementation, and eventually better outcomes. Coordination and sharing will strengthen both groups of actors' ability to save lives in the short term while also building resilience and opportunity for marginalized groups in the future.

In various contexts, including the Middle East, humanitarian actors often face difficulties participating effectively in development discussions due to differing priorities, funding structures, and institutional mandates. Humanitarian agencies primarily focus on immediate, lifesaving interventions during crises, and their involvement in long-term planning can be limited. A common challenge is the fragmentation of efforts between humanitarian and development actors. Urgent needs frequently drive humanitarian responses, and the transition to long-term development may not always be smooth. This disconnect can result in development plans that insufficiently consider complex and evolving situations, hindering the effectiveness of overall aid efforts. Often, humanitarian actors are not invited to the same regional forums as development actors, excluding them from nexus discussions and planning for longer-term WASH interventions. For example, Bastable noted in our interview that, at the Dubai Water Conferences in 2023, many humanitarians were not hosted in the upper-level summits. Regional and global WASH forums must do better to engage humanitarian actors and include them in the same platforms as developmental actors (Bastable 2023). Ultimately, there needs to be more awareness about development interventions among humanitarian and state actors, most notably through regional and global platforms. Whether it be the development actors at fault for inviting humanitarian actors to join nexus planning or the former's unwillingness to do work in the development sphere.

Our survey respondents and interviewees generally agreed that humanitarian actors are not fully included and present in regional and global nexus and long-term development discussions, significantly impacting their ability to think and act with long-term outcomes in mind. Although the humanitarian-development nexus aims to bridge the gap between short-term relief and long-term development strategies, challenges persist in achieving collaboration. As Karine Deniel noted in our interview,

> There exists infighting between humanitarian and development actors (such as UNHCR and UNICEF), for example, over things such as the type of WASH installation that should be installed in the region, and this usually delays the interventions and reduces the impact. Furthermore, the end result is either a WASH installation that addresses either short-term or long-term goals, but not both.
>
> (Deniel 2023)

Cross-collaboration can be easier said than done, and it can be argued that both sectors are complicit in their inability to ensure a more seamless nexus transition. Firstly, addressing infighting between both the humanitarian and development silos, as well as between various other subsectors, is essential for reducing waste. More importantly, it is critical to increase local, national, and international organizations, institutions, and actors' ability to build resilient and sustainable climate change adaptation programing and meet the needs of marginalized groups most impacted by climate change.

One of the most prominent works on the WASH humanitarian-development nexus was developed in 2016 by Mason and Mosello. Although their work focuses primarily on transitional WASH, Mason and Mosello cite structural barriers between humanitarian and development forms that must be overcome to ensure more effective and sustainable services. WASH's development and humanitarian sectors run independently of one another. Our surveys and interviews suggest that this observation has not changed significantly since 2016. Mason and Mosello (2016) argue that the absence of synergy and cooperation results in higher expenses, jeopardizing the longevity and efficiency of interventions and eventually making the impoverished and marginalized population more susceptible to illness and lost socioeconomic prospects. The silos can take many different forms. For example, communities may have their own international dialogue and coordination mechanisms (e.g., SWA versus the GWC); donors may fund development and humanitarian WASH programs in the same countries under different budget lines and for different periods; or implementing agencies may operate independently and place undue strain on their counterparts by ignoring risks, escalating vulnerabilities, or taking inappropriate measures. They highlight at length how the mandates, norms, incentive structures, and procedural processes are built to keep humanitarian and development action separate and siloed (Mason and Mosello 2016). The challenge lies in dismantling existing structures and mandates to enable greater synergy in action. Without such dismantling, achieving effective integration and ensuring successful progress toward climate change adaptation and mitigation become significantly hindered.

Climate Change and Conflict across the Nexus

Addressing the challenging needs of water scarcity and climate change in the Middle East requires a stronger, integrative approach between HDP actors to reduce water waste and increase efficiency in providing safe and sustainable access to water. This is particularly true in fragile states where protracted crises and destabilization are pushing more people into the margins. The World Bank, ICRC, and UNICEF's (2021) recommendations pointed to a greater need for government institutions to lead in developing better data. Still, they pointed to the need for humanitarian and development actors to take greater responsibility for publishing their data for all projects financed through them (World Bank, ICRC, and UNICEF 2021). More specifically, though, there is a need for humanitarian and development actors to work closer together. Humanitarian actors need to recognize and prepare for the very real long-term consequences that come from short-term and short-sighted emergency responses. For example, in protracted crises, humanitarian actors often tap into long-term water sources by bypassing local and national institutional oversight structures, which can often lead to both the increases in the spread of waterborne disease and the further decrease in the ability of local and national government to manage increasingly scarce water resources. Development actors, on the other hand, play a vital role in building reliance before, during, and after a crisis and need to make changes to project cycle programing that allow for greater interconnectedness with local institutions that will ultimately have the responsibility to maintain water infrastructure after the program cycle (World Bank, ICRC, and UNICEF 2021). Even if these changes lead to slower program outcomes, they will have greater long-term success when connected sustainably with resilience mechanisms.

In line with transitional WASH, the focus on climate change and environmental degradation directly impacting people's lives, safety, and well-being cannot be underestimated. This is especially notable in contexts affected by armed conflict, which causes further environmental degradation and hampers people's capacity to recover from shocks and adapt to a changing climate and environment (ICRC 2023). In September 2023, the ICRC Amman Delegation organized the ICRC Regional Resource Network a discussion around the compounding humanitarian consequences of climate change, environmental degradation, and armed conflict in the Near and Middle East, with attention to specific vulnerabilities of displaced people, including transitional programing and WASH services. This discussion unpacked some of the challenges humanitarian and development actors face in boosting action, intending to find a way to operationalize the Humanitarian Development Nexus to increase support to communities affected by conflict in adapting to climate change. Common ground was found on the following points as cited in the ICRC summary overview:

- People's needs are exacerbated by the intersection of environmental degradation, climate change, and armed conflict.
- Humanitarian actors must do their part but have limited capacity to promote the level of adaptation approaches needed in the region.
- Adaptation financing and action are limited in the region.
- To boost action, Humanitarian-Development nexus needs to be promoted, starting with developing a mutual understanding of language and tools, and then find effective collaboration to de-risk and involve multiple actors also in armed conflict settings. Not last recognizing the potential of the private sector.
- More data are needed to demonstrate humanitarian impact of the intersection of climate change and conflict, as well as to document existing adaptation mechanisms and support operations – Displaced people are particularly vulnerable, and adaptation interventions should also be dedicated

to address their needs, without overlooking mobility as a legitimate adaptation mechanism (Darbyshire et al. 2023).

Increasingly, climate change is seen as a protracted and complicated challenge requiring greater long-term coordination and cooperation between humanitarian and development actors. Yet turning these discussions into action is difficult and requires a shift in how practitioners, organizations, and sectors think about and address problems. A typical response often sees organizations divide out the new problem and assign dedicated teams to address just it. This type of response for a challenge as large, complicated, and multidimensional as climate change is impractical and requires a more innovative and integrated path forward.

Yet, as the challenges of addressing climate change continue to grow, there is a risk of further division. This is highlighted in the growing trend of humanitarian and development actors and their funders, separating climate-related programing from other significantly interlinked sectors, like WASH. In contemplating integrating climate change across the HDP nexus, Mena et al. (2022) advocate against establishing climate change as an isolated sector. They argue the imperative to weave climate change perspectives and actions seamlessly into existing HDP streams, rejecting the notion of a separate stand-alone entity. Their research underscores the potential pitfalls of compartmentalizing climate change and argues for a more long-term strategy that integrates climate change and disaster risk reduction into ongoing and future HDP initiatives (Mena et al. 2022).

Looking deeper into the socio-political nature of disasters, Mena et al. (2022) argued further that disasters primarily result from human mismanagement and vulnerability, emphasizing the need for combining climate action and disaster risk reduction with long-term sustainable development. Although some projects and programs have implemented an HDP approach that includes climate-related streams, such as natural resource management, there remains a lack of interventions explicitly integrating all climate change actions. However, this integration would involve a shift away from focusing on climate and weather as potential hazards or influencers toward actions that support preventing, preparing for, mitigating, and managing disasters and conflicts as part of the existing pathways of the HDP nexus. At the moment, neither of these falls within the mandates of either humanitarian or development agencies. This emphasizes a clear gap whereby humanitarian and development organizations are not fully addressing climate change together but operating instead in their own traditional short-term or long-term silos, respectively (Mena et al. 2022). Yet, as we argue here, these findings point to how climate change adaptation is better navigated within the holistic framework of the HDP nexus.

The adaptation required to address climate change, particularly in the Middle East, where climate-related water scarcity and conflict are rising, will require significant local, national, and regional leadership to implement sustainably. Ultimately, the best-case scenario is to build resilient communities where individuals, families, and community institutions can absorb the untraditional shock posed by climate change. This resilience building will require significant local engagement in the design and implementation of adaptation programing as communities have a clearer picture of individual, family, and institutional capabilities and assets, as well as needs and deficiencies. Yet, too much climate-related adaptation programing remains short-sighted and at risk of future shocks (Baxter et al. 2022). Baxter et al. (2022) stated that,

> Ultimately, the humanitarian community must recognise that adaptations are limited and resilience is finite. Climate mitigation must be an utmost priority to ensure a liveable future. Our global interconnectedness leaves humankind uniquely vulnerable to the cascading and compounding impacts of the climate crisis, yet it also offers us transformative new ways to address its threats.
>
> (Baxter et al. 2022, 1562–63)

Although adaptation and resilience programing cannot solve every situation, this reality should not underscore the vital role that humanitarian and development actors can and must play in the successful climate change intervention outcomes across the Middle East. As climate change continues to push people, particularly marginalized groups, beyond their ability to adapt, humanitarian and development actors become vital advocates in helping local and national institutions meet needs. Turnbull, Sterrett, and Hilleboe (2013) argued that long-term prosperity will require linking climate change adaptation and resilience designs at every institutional level and into every sector (Turnbull, Sterrett, and Hilleboe 2013). Shocks are now and will continue to be the norm, and resilience will require movement in both directions across the humanitarian-development nexus. Collaboration and coordination across the HDP nexus cease to be a continuum and become a fluid reality. Turnbull, Sterrett, and Hilleboe (2013) articulated this point further, noting that "As an approach, climate change adaptation is a dynamic process and not an end state," and it becomes the imperative of HDP actors to aid local communities and marginalized groups continually "address current hazards, increased variability and emerging trends; manage risk and uncertainty; and build their capacity to adapt" (Turnbull, Sterrett, and Hilleboe 2013, 4–5).

We recognize that conflict greatly complicates the transition across the HDP nexus. This is particularly true of WASH services, as many current, ongoing conflicts in the Middle East are directly related to climate-related water scarcity. During our interview with Timothy Grieve, he stated, "Climate change is driving and exacerbating conflict, impacting communities, limiting access to services, and disproportionately impacting vulnerable groups" (Grieve 2023b). For example, Yemen faces a severe water crisis due to overpumping of aquifers and limited infrastructure. The ongoing conflict has exacerbated this crisis, resulting in critical shortages of safe drinking water and sanitation services. The protracted conflict in Yemen has damaged water supply systems, contaminated water sources, and displaced millions of people (Al-Salehi 2023; UNICEF 2019a). This humanitarian crisis poses obstacles to the nexus transition in Yemen, whereby the country is moving further away from a phase of stability conducive to sustaining the development of WASH interventions. Many of these conflicts in the Middle East are now connected to upstream water resource management, which further compounds water scarcity downstream. This dynamic can be observed in the climate change-induced droughts and reduced precipitation, which has led to decreased water flow in the Tigris and Euphrates rivers. Turkey's upstream construction of dams and irrigation projects, exacerbated by erratic water availability due to climate change, has heightened tensions with downstream Iraq, which relies heavily on these rivers for its water supply (Giovanis and Ozdamar 2021).

The case of Yemen exemplifies how conflict disrupts access to safe drinking water and sanitation services, hindering the shift from emergency aid to sustainable development. The Tigris-Euphrates River Basin dispute between Turkey and Iraq is another example, demonstrating how climate-induced water stress can escalate tensions and impede cooperation between countries. As a result of water scarcity, conflict constitutes a formidable challenge when implementing a nexus transition between WASH humanitarian programing and developmental programing in the Middle East. In particular, it profoundly affects the transition from short-term to longer-term programing by creating security and access risks for aid workers, damages infrastructures necessary to provide WASH services, displaces vulnerable people, and weakens local and national institutions typically responsible for service provision and maintenance. Overcoming these challenges takes time, and it requires a proactive focus on building long-term resilience and peace in still-accessible populations.

Social Dynamics and Peace

There is a continuous need for a social element of collaboration across the humanitarian and development divide, including peacebuilding. Though not always immediately successful, Tim Grieve highlighted a powerful case study from Dara Syria in 2018 that shows the power of a triple nexus-based response built on collaboration between conflicting communities, noting that

> One example of the triple-nexus and conflict sensitivity, and how WASH projects can improve positive outcomes and reduce negative outcomes in the Middle East, is from a UNICEF project in Dara, Syria in 2018. The UNICEF WASH team was at a crossroads, needing both the power and supply of water between two conflicting communities to successfully

implement their project. One side of the conflict had the power, and the other side had the water resources. A conflict sensitivity assessment was conducted, and out of this came recommendations, actions, and a sense of cooperation. One side did not need to supply power to the other since they already had it but proceeded to do so anyway. This is only one example where the triple-nexus approach is important in the Middle East where there, at times, are many conflicting communities and fragile and informal agreements between one another, on both a smaller and larger scale.

(Grieve 2023b)

These types of solutions, though possible, are difficult as they can be time-consuming and costly. It is difficult for many HDP actors to embed conflict sensitivity into WASH programing as any organization may have dozens of simultaneous ongoing projects all within the Middle East region. Solving every conflict may be impractical, yet adding a conflict sensitivity lens, even as simple as a checklist, can aid in the success of conflict-adjacent programing.

In the end, collaborating across the HDP nexus, with a particular focus on peacebuilding and community participation, is essential to strengthening not only the adaptive capacity of individuals, families, and communities but also the evidence base concerning successful adaptation work, which is particularly lacking in conflict-affected settings. These collaborations can be conducted to empower and strengthen community-level decision-making, which greatly increases the longevity, sustainability, and resilience of WASH and climate-related services. In a recent report by the ICRC and the NRC (2023), they argued that

Humanitarian and peacebuilding actors can help to sharpen the conflict and risk analysis of development and climate actors in order to 'de-risk' and contextualise action in conflict settings, thereby supporting climate adaptation activities that meet the needs of conflict-affected communities. Investment in environmental and climate services, especially in remote areas and informal settlements, will enable anticipatory action and better responses, while empirical studies on successful and unsuccessful adaptation measures in conflict-affected areas are critical to ensure informed decision-making and programming. Drawing on traditional and historic knowledge of climate patterns when systematising climate information will also be necessary in places where hostilities and resource constraints have rendered climate data infrastructure inoperable.

(ICRC and NRC 2023, 56)

Successful collaboration and coordination not only use essential community buy-in but also strengthen social cohesion and trust and, as a result, strengthen the capabilities and resilience of local institutions. As alluded to in our discussion with Tim Grieve, there is a need for improved operational collaboration across the HDP nexus, which can help to harness the complementary nature of both humanitarian and development organizations' mandates and expertise with the critical input and direction of the most vulnerable communities seeking to adapt to a changing climate and environmental degradation (Grieve 2023b).

Additionally, at the organizational level, there is a need to continue strengthening collaboration and coordination with other clusters and areas of responsibility, such as child protection, education, food security, nutrition, health, and shelter, and take on a more strategic approach to inter-cluster themes at global and national levels (UNICEF 2022). In addition, large gaps in preparedness, anticipatory action, the transitions of the wider cluster approach, and humanitarian coordination to local and national leadership cluster's activities are deactivated. The importance of participatory action between all stakeholders cannot be underscored. Greater intentional frameworks are needed to connect local decision-makers with the necessary resources to lead in developing climate-specific WASH services that draw on local strengths while bridging specific weaknesses that stand between sustainability and resilience to climate-specific shocks. These case studies and approaches discussed here underscore the importance of social dynamics in transitional WASH, most notably in post-conflict and crisis contexts in refugee and IDP camp settings.

Camp settings create additional and unique social challenges to transition and tend to incentivize extended short-term interventions that can often last years or decades. For example, the transitional phase in IDP camp settings, like those found in post-conflict Iraq, necessitates a nexus approach underpinned by the closure of camps and the availability and rehabilitation of essential services to facilitate sustainable returns of displaced populations to their home of origin. Agencies like the Camp Coordination and Camp Management Cluster have coordinated assistance for IDPs, aiming to transition camp management to government structures and local authorities to ensure long-term sustainability. With the latest figures indicating that over a million IDPs are still in camps throughout post-conflict Iraq, it's clear that the process of returning IDPs to their original homes is complex and gradual (IOM DTM 2023). Central to this phase is the reinforcement of WASH-related and public health capacities within IDP communities and their areas of origin, exemplified by the handover of these facilities to local management and municipalities. At the same time, international NGOs continue to provide operational training. Risk assessments, mitigation planning, and adherence to a 'Do No Harm' policy are critical during this transition, ensuring that returnees are supported with the necessary information and hygiene supplies. For successful transitions in and out of camp settings, it's crucial to emphasize localization by involving all local stakeholders, such as various levels of local government and their respective departments. Collaboration and coordination with local stakeholders and INGOs are necessary to establish enduring WASH and health solutions and avert further displacement, enabling the restoration of infrastructure and services essential for the reintegration of returnees into post-conflict society (DRC 2021; NRC 2022; Save the Children 2022). These actions weave the social dynamics of peace and community participation into the fabric of transition strategies.

Localization and Sustainability

The concepts of sustainability, localization, and resilience are integrative frameworks that can bring humanitarian and development together under a shared vocabulary and mandate and are best implemented in local and community contexts. When focused on individuals, families, and communities, the difference between humanitarian and development mandates and priorities disintegrates as real-world people with their assets, capabilities, and aspirations emerge. Barakat and Milton (2020) argued that though there has been a long history of commitment to a local focus by humanitarian, development, and government actors, actualization has often been sporadic and requires stronger commitment from all actors involved in climate change response (Barakat and Milton 2020). Yet localization is essential to both the principles of resilience and sustainability. Bypassing local actors may benefit the speed with which humanitarian actors can reach marginalized groups in need, but this comes with the cost of increasing the length of humanitarian need. Even if local institutions are weak in capacity, they often remain operational in some capacity. Tillet et al. (2020) argue that humanitarian actors too often use the mandate to save lives now and the justification to bypass fragile, time-consuming, or effective local and national institutions (Tillett et al. 2020). This not only is a short-sighted strategy but fails to see and take advantage of the opportunity crises that often bring to strengthen or rebuild local institutions that can later lead to resilience, sustainability, and critically a vital transition mechanism as a development entry point and humanitarian exit strategy. The JOF and a strengths-based approach call for greater localization to build meaningful, long-term solutions in transition (Grieve 2023a; Grieve, Panzerbieter, and Rück 2023; Winterford, Rhodes, and Dureau 2023). Effective localization requires humanitarian and development actors to work closely with local leaders and decision-makers to strengthen local institutional capacity, develop handover mechanisms, and enhance collaboration and partnerships that elicit community and beneficiary involvement to achieve long-term solutions.

Climate change is changing the discussion about what constitutes humanitarian and development contexts, challenging traditional divisions in mandates and responsibility. As humanitarian contexts grow in length and scope, past assumptions about short-term disaster response and long-term development responsibility push for new norms and understandings. The New Way of Working and the development of the HDP nexus are the latest push for greater collaboration and connectivity between humanitarian and development actors (Barakat and Milton 2020). A recent UNICEF and GWP publication stated that

> While a rapid response to disasters is critical, there are enormous oppor-
> tunities to integrate climate resilience into the design, siting, construction,
> delivery and monitoring of humanitarian interventions, as well as oppor-
> tunities for mitigation. The increasing application of Environmental and
> Social Safeguard systems into programmes, many of which include stand-
> ards relating to community health, resource efficiency and climate change,
> will also greatly help the periodic assessment of climate and water scarcity
> risks in humanitarian programming.
>
> (UNICEF and GWP 2022, 5)

Achieving this integration requires a unified response, bringing together the strength of actors on all sides of the HDP nexus. Resilience-building and sustainability frameworks provide a critical shared vocabulary and mandate that can help bridge traditionally siloed responses together under a shared mandate to help marginalized groups across the Middle East address and respond to the current and future pressures of climate change. The World Bank, ICRC, and UNICEF noted in 2021 that this type of response will require humanitarian actors to "complement their emergency response man-dates with preventive approaches that support the business continuity of WSS service providers" by increasing their focus on long-term outcomes in addi-tion to traditional, short-term, lifesaving programing (World Bank, ICRC, and UNICEF 2021, 22). On the other hand, governments and traditional develop-ment actors need to reprioritize program designs on principles of resilience and sustainability. The same World Bank, ICRC, and UNICEF report high-lighted that many governments and development organizations had paid too little attention to resilience building pre-crisis and failed to adequately prepare for even the most likely of disasters, often relying on one source to meet all of their water or power needs with little to no contingency plan if these sources fail (World Bank, ICRC, and UNICEF 2021). Continuing work as normal is no longer an option. Integration across the HDP nexus is now the only way to address the region's complexity of climate change in the Middle East.

Technical Challenges across the Nexus

Recognizing the intricate dynamics at play, the conversation on enhancing WASH services amidst the challenges posed by climate change necessitates a broader, more inclusive approach. As we navigate the complexities of WASH services in regions affected by climate change, it becomes increasingly clear that addressing these challenges goes beyond mere technical solutions. The insights from Winterford, Rhodes, and Dureau (2023) underscore the sig-nificance of understanding WASH issues through a lens that appreciates the interplay of social, economic, political, and cultural factors. By acknowledging the role of social accountability systems, power dynamics, gender relations, and organizational responsibilities, we can see the potential for a strengths-based approach that leverages local capacities and knowledge. This perspective not only aims to enhance service delivery but also seeks to harmonize WASH initiatives with the broader aspects of human life, thereby fostering a more integrated and sustainable response to the adversities brought forth by climate change. The call for a strengths-based approach in WASH programing encourages a shift from a deficit-focused perspective to one that values and builds upon the existing strengths within communities. Such an approach not only promises to address the immediate needs related to WASH but also aligns with the overarching goals of sustainable development and resilience against climate-related challenges (Winterford, Rhodes, and Dureau 2023, 227).

However, as the WASH sector strives for localization, integration, resilience, and sustainability amidst climate change and the nexus transition, practitioners must develop new technical capacities to meet these evolving challenges effectively. This requires developing more diverse and advanced skill sets and mindsets, presenting additional challenges for organizations transitioning across the nexus. Esmaeil Ibrahim stated in an interview that "there is also a lack of technical knowledge, especially at the local level, to enable resilience and sustainability" (Ibrahim 2023). In this shifting landscape, it's crucial to address the shortage of technical knowledge, particularly at the local level, which is fundamental to achieving resilience and sustainability in the WASH sector. As Esmail aptly points out, this lack of expertise at the grassroots level can impede the implementation of effective strategies and solutions. To overcome this hurdle, there's a growing need for capacity-building efforts, knowledge transfer, and the cultivation of local expertise. Empowering local actors and communities with the skills and knowledge required to address WASH challenges in the face of climate change and the nexus transition is essential for the sector's progress. Yet, we must avoid a deficit perspective that relies on external evaluations to determine which skills must be filled. Instead, local institutional decision-makers should be the driving force behind identifying local skill assets and potential technical needs that can be addressed (Winterford, Rhodes, and Dureau 2023). Furthermore, fostering collaboration among local, national, and regional WASH practitioners, governmental bodies, and relevant stakeholders can help bridge the knowledge gap and encourage the development of innovative, sustainable, and resilient approaches to tackle these complex challenges effectively.

Yet, as Tim Grieve argued in our discussion, addressing these gaps requires many training sessions to be completed, on top of trying to deliver on WASH projects, raising issues around efficiency and timely project implementation (Grieve 2023b). As such, skills building would also require mandates, dedicated time, and investment toward nexus-thinking from humanitarian and development actors for all WASH focal points within local and national institutions and government bodies. Nexus-thinking requires training qualified and experienced WASH focal points at the regional and country levels who can address WASH needs along the continuum of challenges from the onset of a humanitarian emergency through a protracted crisis to the post-conflict recovery. Increased local capacity will ensure long-term and sustainable WASH outcomes that can make sustainable transitions across the HDP nexus.

In addition, recognizing the interconnected and integrated nature of WASH with various sectors, such as health, nutrition, shelter, education, child protection, and livelihoods, is essential for developing comprehensive and sustainable solutions. By adopting a nexus approach, donors, implementing organizations, and local actors can foster collaboration and integration across disciplines, bridging the gap between technical expertise and broader development goals. This holistic approach enables the identification of synergies, shared resources, and common objectives, which can contribute to a more effective allocation of resources and targeted capacity-building initiatives. Moreover, integrated WASH programing and nexus and resilience thinking encourage cross-learning and knowledge sharing, allowing professionals with diverse backgrounds to collaborate and contribute their expertise to collectively tackle the complex challenges that climate change will pose across the Middle East.

A significant challenge to developing and maintaining critically skilled capabilities is brain drain, whereby the retention of highly qualified staff with high levels of skills in addressing climate change and WASH needs is not financially incentivized to remain in their country of origin within the Middle East. Brain drain poses a significant technical challenge due to its impact on human resources and expertise across multiple sectors. The emigration of highly skilled professionals, including engineers, technicians, and researchers, from the region to seek better economic opportunities abroad results in a loss of critical knowledge and capacity essential for effective adaptation and mitigation planning, implementation, and management. This brain drain hampers the region's ability to develop and sustain innovative solutions, manage complex water systems, and adapt to changing environmental and climate conditions. The departure of skilled personnel also contributes to a shortage of trained workforce capable of addressing evolving challenges, such as water scarcity and sanitation issues, thereby impeding the region's progress in achieving adequate and sustainable WASH outcomes.

Yet, not all of our interviewees agreed on the true nature of critical skill gaps. In one such case, Andy Bastable argued the contrary, stating, "There is no technical gap. The real issue is an unwillingness to understand the local mechanisms and involve them from the beginning" (Bastable 2023). Greater localization, therefore, becomes critical to successfully addressing technical shortages yet also represents a critical gap preventing long-term sustainability and resilience in WASH services. Andy Bastable noted that in some instances, companies are contracted in the absence of the government, which does not have a solid track record either. He stated that,

> In the Middle East, many water treatment plants are not working anymore due to the lack of community-managed systems. In Iraq, compact treatment units do not work anymore, need maintenance checks, and lack sustainability due to a lack of community management. Also, in Yemen, a few desalination plants are no longer working for the same reasons.
>
> (Bastable 2023)

Too many climate-related WASH services are either inadequately tied to local management systems or bypass local or national institutions altogether, leading to short-term solutions that become derelict in a few short years. The GWC echoes the call for stronger, long-term coordination with local institutions and decision-makers that strengthen capacity development and learning and introduce more diverse approaches to a broader audience that are more contextualized and localized and include local and national actors (UNICEF 2022).

Essentially, localization is the solution that enables a long-term, sustainable, and resilient transition across the HDP nexus. The lack of skills and capacity, exacerbated by brain drain and a deficiency in community-led programing, underscores the urgency of addressing these technical gaps. Climate change's complex and interconnected nature across many different sectors necessitates a paradigm shift towards nexus thinking, where collaboration, integration, and localization are prioritized, even from the onset of conflict. This paradigm shift allows for asset, strength-based thinking over typical solutions geared around filling perceived deficits and empowers individuals, families, and communities to become the diviners of their solutions. To effectively confront the multifaceted challenges of the region along the HDP nexus transition, humanitarian and development actors must not only just bridge technical gaps but also adopt nexus-thinking that fosters a resilient, skilled, and community-oriented approach that aligns with the evolving natural climate realities and the interconnected needs of the Middle East.

5   Resilience, Sustainability, and the HDP Nexus

The Middle East faces a complicated and uncertain future. Conflict and water scarcity appear to be two of the most significant constants in the region, and climate change is only proving to make these challenges larger and more difficult to address. Humanitarian crises are shaped by the complex interplay of financial, institutional, environmental, technical, and social challenges and require a paradigm shift toward holistic and adaptive strategies that acknowledge and address these challenges in the Middle East. To achieve sustainable and resilient WASH services, mitigation and adaptation efforts will require critical changes to financial, institutional, environmental, technical, and social practices and structures, enhancing greater collaboration, coordination, and integration across the HDP nexus. The interconnectedness of the five pillars of the sustainable WASH model highlights the need for a multisectoral approach to climate change and water scarcity between sectors and sub-sectors and across the HDP nexus. This perspective underscores a shift from reactive, short-term solutions to proactive, inclusive, and nexus-driven strategies prioritizing long-term sustainability and resilience.

Climate change has a tremendous, yet unequal, impact on everyone's daily experience across the Middle East region. Addressing the complexities of climate change requires tailored mitigation and adaptation efforts that ease the unique climate-related pressures on sectors like agriculture, food security, livelihoods, health, nutrition, shelter, education, child protection, and resource management, including energy, water, and WASH services. Mitigation efforts are often short-term, reactionary responses after shocks are felt at the individual, local, national, or regional levels. On the other hand, adaptation efforts seek to build sustainability, resiliency, and redundancy into climate response by focusing on the long-term capabilities of individuals, families, communities, and nations to absorb future shocks and stressors while maintaining and growing essential local and national infrastructure. As national and regional crises grow in length and scope, the divisions between short-term and long-term climate response necessitate reevaluating the traditional division between humanitarian, development, and peace mandates and priorities. There is a greater need now than ever for short-term humanitarian action to be

DOI: 10.4324/9781003436706-5

coordinated and well-connected with long-term development outcomes. This long-term vision must be designed and implemented by local and national institutions that hold the mandates for long-term peace. This will require a fundamental shift in how HDP actors design, implement, coordinate, and achieve desired outcomes. Institutional bodies overseeing climate response may still lack critical capacity, requiring the assistance of international humanitarian and development resources and expertise. Despite these weaknesses, building effective and intentional transition points across the HDP nexus is essential to long-term climate adaptation and stabilization.

There is a critical need for an integrated approach in transitioning climate-related initiatives from emergency humanitarian responses to sustainable, long-term development strategies in the Middle East. Success will require rethinking traditional practices that foster effective collaboration across the HDP nexus, emphasizing localization and sustainable practices. This shift to a holistic nexus approach, integrating resilience, sustainability, and community empowerment frameworks to address climate challenges, underscores the necessity of greater collaboration, innovation, and localization to reach a more sustainable, resilient, and equitable future. The multisectoral nature of climate change and its impacts make its risks particularly complex for mitigation and adaptation efforts. The unpreparedness of various sectors, particularly WASH, to effectively address climate change consequences during times of stability leads to heightened vulnerabilities, especially among the most at-risk populations. Of particular focus are the intricate relationships between institutional weaknesses and the challenges posed by climate change in the Middle East, which profoundly affects the region's ability to develop and sustain effective WASH services. We argue here for robust institutional frameworks and climate change-focused policies, emphasizing a coordinated, proactive approach that increases the capabilities of local, national, and international actors to navigate the complexities of climate change.

This research started with recognizing the complex linkages between WASH services and the growing climate change crisis in the Middle East. Our findings, however, show the multisectoral nature of climate change and WASH and, therefore, require a multisectoral response larger than WASH could ever provide alone. We are noting a growing trend that deemphasizes WASH as an individual sector and highlights how other sectors, like health and climate change, are beginning to absorb a greater portion of traditional WASH services. We generally agree with this approach. As we have argued here before, WASH, as a sector, is deeply linked with sectors like health, nutrition, shelter, education, and child protection and therefore, deserves to be a significant contributing factor to their response, yet WASH services may be of more benefit to at-risk and marginalized groups through its absorption into these various sectors than by standing apart. Specifically in the case of climate change, WASH services and WASH programatic designs are essential to all mitigation and adaptation efforts if long-term sustainability and resiliency are to be achieved. In this book, we have drawn heavily on the Sustainable WASH Model, also known as the FIETS model (WASH Alliance International 2021), with its focus on the financial, institutional, environmental, technical, and social aspects required for effective WASH services, along with the Joint Operational Framework (JOF) (Grieve 2023a; Grieve, Panzerbieter, and Rück 2023) and a Strengths-based Approach (Winterford, Rhodes, and Dureau 2023) to development. Though initially designed with WASH services in mind, we argue that the Sustainable WASH Model and the JOF are as applicable to WASH services specifically as they are to climate change broadly. In this chapter, we also draw on three new case studies from Yemen, Iraq, and Palestine as examples of progress in the right direction. Though these case studies are not comprehensive, they offer new insights into ongoing discussions around integrative HDP response. The full case studies are in Annexes 3, 4, and 5.

A Nexus Mindset

For humanitarian and development practitioners, the growing pressure of increasing numbers of protracted crises, coupled with ever-scarcer resources to not only mitigate the impact disasters have on vulnerable populations but to build long-term resilience, is their single greatest challenge. As resources and capabilities become increasingly more difficult to manage, it incentivizes HDP actors, who are generally risk-averse, to rely on simpler, predominantly outcome-focused, and cost-effective pathways. Our argument here is that how outcomes are achieved is far more impactful in the lives of marginalized and vulnerable populations. Even if those outcomes must be achieved through slower and more complicated processes. This new way of thinking is the essential tool HDP actors must develop if the increasingly complex problems of climate change are to be addressed sustainably, resiliently, and meaningfully. Growing complexity, particularly around climate change, drives a shift from traditional ease of thinking, designing, and implementing HDP programing toward an integrated, nexus mindset capable of empowering, innovating, and sustaining long-term climate adaptation and resilience.

To successfully transition from siloed humanitarian and development approaches to a more integrated nexus model, there is a pressing need for concerted efforts and collaboration at multiple levels. The study by Mason and Mosello (2016), which looked specifically at transitional WASH, concludes that addressing the deep-rooted challenges requires a proactive approach from WASH professionals at the operational level, where practical compromises can be forged. To facilitate this transition, a leader in the WASH sector at the international level, such as UNICEF, plays a pivotal role in facilitating agreements on a limited set of shared priorities and core ways of working between development and humanitarian WASH stakeholders in pilot countries. Simultaneously, donors providing funds for WASH programs should further prioritize flexibility and longer-term initiatives to overcome underlying incentives hindering complementarity and enhancing the effectiveness of responses in protracted crisis contexts. A 'Champions Group' currently led by the GWC nexus group should be prioritized to a greater extent to actively work on incentivizing collaboration across siloes. Additionally, addressing cultural and systemic barriers beyond the WASH sector necessitates the establishment of cross-sector initiatives by secretariats of WASH partnership and coordination structures at global, regional, and country levels. By fostering complementarity and emphasizing the role of local actors and leaders, these initiatives can contribute to the resilience and well-being of marginalized communities, reducing vulnerability to diseases and missed socioeconomic opportunities (Mason and Mosello 2016; NRC 2023). We argue here that these conclusions are equally relevant now and apply just as well to broader discussions around the humanitarian-development nexus as they do to the WASH sector.

The Organization for Economic Cooperation and Development (OECD), at the World Humanitarian Summit 2016, proposed key solutions to enhance the transition between humanitarian and development efforts. They argued that the conventional view of crises as isolated incidents along the development path has evolved, recognizing that crises are intricate and prolonged and require more than immediate lifesaving aid. The existing donor aid structure often segregates humanitarian and development teams, leading to disjointed approaches with separate tools, funding cycles, and decision-making processes. The OECD advocates for a paradigm shift to address this, emphasizing coherence in aid strategies (OECD 2017). The OECD's approach to bridging the humanitarian and development divide centers on achieving coherence by aligning humanitarian and development objectives, analyzing the comparative advantages of financial instruments, and promoting collaboration through partnerships. This involves a shared, risk-informed context analysis, mobilizing aid instruments based on their strengths and fostering political leadership to overcome institutional barriers. Coherence is not about integrating humanitarian assistance into a political agenda but ensuring that humanitarian principles are maintained. The OECD asserts that coherence is essential for addressing complex crises effectively and efficiently, promoting sustainability, and accelerating progress toward Sustainable Development Goals (SDGs) in challenging environments (OECD 2017).

At the World Humanitarian Summit 2016, the New Way of Working (NWOW) was also developed, aiming to pursue collective outcomes collaboratively between humanitarian and development actors over multiple years, leveraging diverse actors' comparative advantages (OCHA 2017; United Nations 2017). The NWOW calls for collaborative action between humanitarian and development actors, leveraging their respective strengths to achieve 'collective outcomes' that aim to reduce long-term needs, risks, and vulnerabilities over several years. The NWOW emerged as a response to the escalating volume, cost, and duration of humanitarian assistance over the past decade, primarily driven by the prolonged nature of crises and insufficient development initiatives in vulnerable contexts. This trend underscored the pressing need for improved connectivity between humanitarian and development efforts. This approach gained widespread support and commitment at the World Humanitarian Summit, signifying a concerted effort to bridge the divide between humanitarian and development endeavors.

At the core of NWOW is the concept of "collective outcomes," encapsulated in a Commitment to Action signed by the Secretary-General and nine UN Principals at the WHS, with endorsement from the World Bank and the International Organization for Migration (IOM). These collective outcomes represent concrete and measurable results that humanitarian, development, and other relevant actors aspire to achieve jointly over a multi-year period. NWOW provides a practical framework for translating the shared vision outlined in the 2030 Agenda and the SDGs into actionable strategies aiming to support the most vulnerable populations and ensure that no one is left behind in future development efforts (United Nations 2017).

Effective collaboration between humanitarian and development actors is vital for successfully transitioning in the humanitarian-development nexus. Despite differing mandates, working together is the only path that ensures a seamless transition from emergency response to long-term development and leverages the strengths of both sectors. This collaboration enhances contextual understanding, resource efficiency, and inclusive decision-making, ultimately contributing to more resilient communities, improved funding mechanisms, and aligned policies. By overcoming challenges and jointly addressing immediate needs and underlying issues, humanitarian and development actors can create comprehensive, sustainable, and locally informed solutions for communities in crisis and transition settings.

There is an urgent need for a collaborative and integrated approach to transition from the traditional siloed methods of humanitarian and development work to a more cohesive nexus model. Mason and Mosello (2016), and a recent NRC (2023) report, emphasize the importance of proactive engagement from WASH professionals and the critical role of organizations like UNICEF in mediating agreements between humanitarian and development stakeholders. They also call for donors to support flexible, long-term initiatives and suggest fostering champions to focus more on cross-sector collaboration (Mason and Mosello 2016, NRC 2023). Additionally, the OECD (2017), at the World Humanitarian Summit, advocates for a paradigm shift towards coherence in aid strategies, recognizing the complexity of crises and the necessity for a unified approach to humanitarian and development efforts. The NWOW introduced at the same summit (OCHA 2017; United Nations 2017) aims for collective outcomes through collaborative efforts across humanitarian and development sectors to address long-term needs, risks, and vulnerabilities effectively. A decade later, these recommendations still stand as they collectively stress the importance of collaboration, flexibility, and a shift in perspective to effectively address the challenges posed by complex crises, which are key to sustainable development and resilience among vulnerable communities in transition.

A Strengths-based Approach

A Strengths-based Approach to climate change response provides a critical shift in the optics of HDP actors. It is a critical component of an effective nexus mentality. A Strengths-based Approach uses appreciative and positive language that is asset-focused as it helps isolate local communities' strengths, resources, and capabilities in addressing complex problems like climate change. Critically, a Strengths-based Approach puts local stakeholders and decision-makers in the driver's seat of change and uses HDP actors as facilitators to fill gaps as capacities are strengthened and enhanced temporarily. What makes a Strengths-based Approach not unique but effective, argues Winterford, Rhodes, and Dureau, is that it,

> First, strengths-based approaches are orientated towards creating change, not simply focused on revealing 'positives' within current situations, The change process seeks to create alternative futures by drawing on what is working in the present, by revealing and amplifying these in order to create change, It is the 'appreciative stance' which harnesses the strengths of the past and potential of the future for change. Second, change is not simply about fixing the present, but is orientated towards transformative change, that is, new situations, contexts in which individuals, groups, or organizations can operate. The future orientation and focus on transformative change are important distinctions of the change process in a strengths-based approach.
>
> (Winterford, Rhodes, and Dureau 2023, 69)

We argue that the critical power behind a Strengths-based Approach is its emphasis on correct foundations for transformational change, focused on the correct means for success over simple, linear, and cost-effective outcomes. The most effective way to reach long-term outcomes is through empowering individuals, families, and communities to positively use their capabilities to confront complex problems in sustainable and resilience-enhancing ways. Often, appropriate and tailored facilitation by HDP actors is all that is needed to start effective and transformational change.

Yet, it takes time to convince HDP workers and their respective organizations of the essential positive merits of a Strengths-based Approach. Winterford, Rhodes, and Dureau argued that from their experience advocating and using strengths-based approaches, globally that

> Aid agencies can take some convincing to realize that this is key to the process of sustainable development. The irony is that the greater the investment in early local-level strengths-based facilitation, the

shorter the overall life of the project will be and the more sustainable the benefits.

(Winterford, Rhodes, and Dureau 2023, 137–38)

In a world of protracted crises, many of which are only growing in scope and cost, building early foundations of local engagement can pay off with divides down the line and, in many cases, are more effective avenues in creating sustainability, resilience, and, critically, exit strategies for HDP actors. Yet the mind frame shift toward a strengths-based version of nexus-thinking can also disorientate. A Strengths-based Approach, emphasizing facilitation, takes the control over outcomes away from HDP actors, which can be uncomfortable for humanitarian and development workers who are used to full ownership over the design, implementation monitoring, evaluation, and learning process. Over time, the facilitation process can become natural and preferred as local-led change has a greater potential to create stronger, more resilient outcomes than traditional internationally led initiatives (Winterford, Rhodes, and Dureau 2023).

Developing a nexus mindset remains difficult for HDP workers. Persistent and robust silos divide humanitarian and development actions into sectors that, due to fundings structures and professionalization, make dismantling and redefining madness increasingly difficult (Winterford, Rhodes, and Dureau 2023). Yet nexus discussions and the NWOW challenge these structures and provide space for further shifts from traditional deficit responses to HDP programing. Taking this argument further, Winterford, Rhodes, and Dureau stated that

> The current problem-based approach which underpins international development practice is simply not suitable for achieving successful collaboration for change, across cultures and in every corner of the world. Successful collaboration requires trust and respect. Trust and respect require recognition of everyone's agency and potential, understanding of diverse cultural values, localization of goal-setting and decision-making, and decolonization. A strengths-based approach contributes to all of these layers.

(Winterford, Rhodes, and Dureau 2023, 233)

Therefore, a Strengths-based Approach to climate change challenges in the Middle East becomes the foundation of collaboration and coordination structures prioritizing local lead sustainability and resilience. The development of a strengths-based, nexus-focused mindset is a critically essential step for HDP practitioners and organizations if the complexities of climate change adaptation and mitigation efforts are going to be successful.

The Joint Operational Framework

The JOF, developed by collaboration with the Global WASH Cluster, UNICEF, SWA, and German WASH Network, when integrated with critical concepts from a Strengths-based Approach, becomes a significant model for coordinating and building effective collaboration between local, national, and international actors. Although a Strengths-based Approach has inherent disagreements with the JOF's reliance on needs assessments, risk evaluations, and general focus on vulnerabilities, which are inherently deficit-based, the JOF provides a valuable framework for effective, local, national, and international collaboration. It is not hard to conceive of a variation of the JOF that is reframed to be asset-focused, emphasizing local skills, opportunities, and strengths while utilizing all the same collaboration structures. Integration across the HDP nexus is enhanced by both the JOF and the strengths-based approach, as both refocus actors on local stakeholders, leadership, and decision-makers while providing access to international resources and capacities that might temporarily be absent. It is critical to state that a Strengths-based Approach does not refute that deficits do not exist but simply focuses on identifying and filling them on local stakeholders. Yet, by utilizing a Strengths-based Approach, facilitators can significantly increase local trust and buy-in and strengthen relationships in, between, and around communities, which are all good HDP outcomes (Winterford, Rhodes, and Dureau 2023). Although the JOF is specifically WASH-focused, the coordination mechanisms align well with broader contextual problems like climate change and crisis management in general.

The JOF provides an essential framework for effectively actualizing a nexus mind frame. Mason and Mosello (2016) argue that humanitarian and development actors' mandates, norms, incentive structures, and procedural processes are built to separate themselves into their own silos. Effective integration requires dismantling these structures and mandates to allow for greater synergy in action, which is the only way to ensure that progress toward climate change adaptation and mitigation will succeed (Mason and Mosello 2016). In adopting the nexus mindset, it is imperative to move beyond rhetoric, enhance inclusivity by listening to the affected populations, and secure flexible, multi-year funding to align with the dynamic nature of crises in the region (ICVA 2022; Land and Hauck 2022). The JOF seeks to do this by building a collaborative framework that allows for changes in siloed structures to allow more flexibility, which is necessary to address protracted crises like climate change.

Integration between and across the HDP nexus is critical to responding to protracted crises and requires humanitarian and development actors to enter crisis both early and vitally together. Long-term outcomes require long-term development actors' input early in a crisis and must build on a foundation of local engagement and trust, as argued here. Transitions from humanitarian to development actors are difficult and require intentional considerations. Grieve, Panzerbieter, and Rück (2023) articulated four critical features of effective integration and coordination between humanitarian and development actors, arguing for

1 Humanitarian actors, in consultation with development actors, to focus on sustainability from the start of an emergency response, building on existing water resource/WASH systems and markets.

2 Both humanitarian and development actors to develop policies supporting each crisis phase with associated guidance on triggering, standards, data sharing, and associated accountabilities for all stakeholders.

3 Development actors to support the integration of humanitarian actors' input into long-term assessment, policy, planning, and resourcing processes.

4 Humanitarian and development actors to collaboratively transform monitoring systems to include relevant "real-time" indicators that incorporate resilience, conflict sensitivity, and peacebuilding capacities (Grieve 2023a, 40–41; Grieve, Panzerbieter, and Rück 2023, 8).

The earlier humanitarian and development actors begin working together, the greater and more effective the transition between mandates is and increases the potential for long-term sustainability reliance while also reducing duplication and the cost on donors.

In the Middle East, climate change is a water problem that exacerbates conflict and violence over scarce resources. Therefore, integrative solutions that rely on multisectoral coordination mechanisms between various actors in WASH, resource management, water security and environmental protections, and agricultural management will also impact the long-term sustainability of climate change mitigation and adaptation programing in the region. The focus should be on shared learning and flexibility, allowing for greater control in responding to current and future shocks. Grieve, Panzerbieter, and Rück (2023) argue in the JOF for integrative adaptive management with the goal of "creat[ing] an environment of intentional learning and flexible project and activity design." The authors of the JOF argued that this "requires minimizing the obstacles to modifying programming and creating incentives for adaptive management" (Grieve 2023a, 40; Grieve, Panzerbieter, and Rück 2023, 8). How humanitarian and development actors can bring their joint comparative advantages to protracted crises.

Forming these coordination efforts requires international advocacy by all actors across the HDP nexus toward clearer collaboration between all potential partners with a stronger emphasis on clearer transitionary strategies. Specifically, humanitarian and development actors must be invited to the same national, regional, and international forums. Too often, these discussions, forums, and meetings are held for one group or the other. This also applies to coordination meetings, where development actors should be invited to and participate from the very beginning of a crisis, even if development interventions cannot be utilized for quite some time. This creates stronger avenues for shared data, coordinated transitions, and a clearer exit strategy for humanitarian actors when the time comes.

Localization

Localization has been a key theme throughout our discussion. Yet, shifting the mindset of humanitarian and development workers from program designers and implementors to a new role as facilitators aimed at aiding local decision-makers in designing and implementing their solutions to complex problems is essential to building long-term, sustainable, and resilient outcomes. In a 2021 report by the World Bank, ICRC, and UNICEF, they argued that "a major driver of the decline in services during protracted crisis has been the lack of attention paid by development partners to resilience building before crisis" (World Bank, ICRC, and UNICEF 2021, 62). Sustainability and resilience outcomes are not only built on the foundation of strong integration with local actors but critically require a greater focus by HDP actors on long-term consideration of resilience and what future shocks will look like. The ICRC and NRC noted that "people and communities on the front lines of climate change – including displaced people – are often in the best position to identify the most pressing risks and issues and to contribute to finding solutions" (ICRC and NRC 2023, 55). For HDP actors, building local trust and expanding local networks are critical to finding and building the most effective climate adaptation and mitigation outcomes.

However, localization brings unique challenges that can be difficult for HDP actors to navigate effectively, especially when addressing climate change, a complex, multisectoral problem requiring commitment from a diverse group of actors, all with unique priorities and perspectives. It is common to find local stakeholders and decision-makers in one sector or institutions that are fully onboard with desired adaptation programing and then to find another sector or institution that has a different, conflicting perspective or set of priorities that shuts down progress (Winterford, Rhodes, and Dureau, 2023). Achieving localization requires HDP actors to engage with diverse perspectives and, as facilitators, to provide avenues that lead to common ground and a shared vision for the future (Winterford, Rhodes, and Dureau, 2023). It must be stated here that even at the local level, not all individuals, families, and communities share the same degree of power, capacity, or ability to engage in decision-making. Although we argue that local decision-makers should drive climate change mitigation and adaptation efforts, HDP actors ensure that all local groups are part of the decision-making process (Winterford, Rhodes, and Dureau 2023).

UNICEF's case study on Greening in Schools and Health Care Facilities (HCF) in Iraq (See Annex 4) highlights key successes around community engagement in effective climate-related interventions and localization efforts. This program, which implements green, solar-powered water systems, signifies a multifaceted approach to capacity building and sustainable water management. By ensuring these institutions met the necessary water and energy requirements, UNICEF facilitated the installation of renewable energy-powered water systems and laid the foundation for an educational program emphasizing the importance of water conservation, renewable energy, and environmental stewardship among students and community members. Additionally, this program provides hands-on experience in monitoring and

maintaining these systems, fostering a sense of local responsibility and innovation. This program goes beyond simple infrastructure improvement by
embedding environmental education to empower children and their communities by fostering a culture of sustainability and resilience. Through these
efforts, UNICEF addresses immediate water scarcity issues while inspiring
a future generation to adopt and learn to maintain sustainable practices and
technologies, contributing to long-term resilience and water security in Iraq.

In the Yemen case study (see Annex 3), the IOM highlights an ongoing
program focused on monitoring groundwater levels, highlighting the applicability of localization by humanitarian organizations in fragile, conflict-prone
contexts. In Yemen, the IOM has actively engaged in the capacity building
of authorities and water humanitarian stakeholders to address the severe
water scarcity and the effects of the ongoing conflict on water resources. The
IOM aims to sustainably enhance local expertise in managing scarce water
resources by providing specialized training in groundwater monitoring for
WASH actors. This initiative includes the set-up, operation, and data analysis from groundwater monitoring equipment, which is crucial for effectively
understanding and managing groundwater resources. Additionally, the IOM's
efforts to educate agricultural users on the importance of efficient water
use and implementing sustainable irrigation methods are vital for reducing
groundwater pollution and overexploitation. The exploration of natural and
artificial aquifer recharge techniques during the brief rainy season represents
an innovative approach to augmenting water supplies, further emphasizing
the IOM's commitment to sustainable water management in Yemen's challenging context.

Adaptation, Sustainability, and Resilience

A nexus mindset creates greater potential for moving past simple shock and
response strategies to climate change and toward approaches that allow communities to anticipate, adapt, and thrive in a future of uncertainty. Adaptation
to climate change requires unique skills and foresight that are not intuitively
learned or taught through traditional systems and, two, critically lacking in
marginalized populations most at risk of climate change's current and future
risks. Werners et al. (2021) created the term 'climate-resilient development
pathways' and noted that resilience building "reorients the climate challenge
from delivering on climate targets to facilitating long-term sustainable development. In this context, resilience, pathways, and transformation can be considered boundary concepts to consolidate climate action and development
decisions towards long-term sustainable development" (Werners et al. 2021,
174). Effective orientation on sustainable and resilient adaptation requires
complex local conversations around economic, environmental, and livelihood
impacts and how best to prioritize resource management and program design.
Only local stakeholders and decision-makers are adequately placed to make

this complex decision. Yet, conceptual frameworks built around sustainability and resilience are critical common ground for forming and conceptualizing progress.

A recent IMF report noted that a broad set of policy recommendations is needed throughout the Middle East region to address climate change comprehensively, including emission reduction, renewable energy, water management, resilient infrastructure, and green investment (Duenwald et al. 2022). Middle Eastern governments must adopt a comprehensive and multifaceted approach to address climate change. This would involve setting and enforcing stringent emission standards to limit greenhouse gases from key sectors like industry, transportation, and energy. Simultaneously, there should be a strong push towards renewable energy, capitalizing on the region's abundant solar and wind resources and incentives to foster their development and use. Water resource management is crucial, given the region's water scarcity; this entails policies for sustainable use and conservation efforts (Duenwald et al. 2022).

Building climate-resilient infrastructure is another critical step that ensures that new constructions can withstand extreme weather events. Raising public awareness and education about climate change impacts and mitigation strategies is essential for cultivating a community-wide understanding and response to environmental challenges. Governments must support sustainable agriculture practices, such as drip irrigation and crop rotation, to maintain food security in changing climate conditions. Lastly, promoting green investments through financial incentives can encourage businesses and projects to contribute to environmental sustainability, aligning economic development with ecological preservation. By integrating these strategies into a cohesive policy framework, Middle Eastern governments can significantly contribute to global climate change mitigation efforts and secure their regions' long-term sustainability and resilience. Ultimately, there needs to be a responsibility to maintain ownership, commitment, and accountability on the government's side in taking action and imposing the regulations necessary to combat climate change in place of other actors (Duenwald et al. 2022).

Creating forward progress, which includes effective water adaption, sustainability, and resilience, requires focusing on how climate change will create future shocks and a contextually tailored plan on how current water systems can function despite those stresses. This adaptation and reliance require adopting new technologies that reduce water usage and reconfiguring systems, institutions, and decision-making models better suited to meet future climate change shocks (Winterford, Rhodes, and Dureau 2023). These changes can be uncomfortable and, in areas of protracted crisis, can be difficult to initiate. New partnerships between HDP actors are essential to creating effective long-term sustainability and resilience. A recent report by the NRC (2023) noted that traditional humanitarian action saves lives yet questioned the effectiveness of spending millions of dollars on humanitarian interventions in protracted contexts that have not been designed to increase long-term adaptation, sustainability, or resilience in the lives of the people served (NRC 2023). Creating effective adaptation, sustainability, and resilience must be started at the very beginning of a crisis to be effective, and it requires a fundamental shift in the way traditional humanitarian and development actions have traditionally been divided.

In the Iraq case study (see Annex 4), the adoption of solar power, smart water technologies, and the development of Enterprise Performance Management (EPM) software by UNICEF represents a significant stride toward integrating innovation and technology within the framework of climate-resilient WASH infrastructure. These initiatives exemplify a proactive approach to climate adaptation, ensuring that vital WASH services remain uninterrupted in the face of environmental pressure and climate variability. This strategic response secures essential health and educational services and demonstrates the benefits of renewable energy and efficient water use to the broader community. By focusing on sustainable technology in educational and healthcare settings, particularly children's spaces, UNICEF demonstrates the transformative potential of innovation in building resilience, reducing dependency on non-renewable energy sources, and instilling a culture of environmental consciousness within communities. This example highlights critical intentionality in climate-related adaptation, sustainability, and resilience program design. This approach reduces dependency on non-renewable energy sources, reduces greenhouse gas emissions, and promotes longer-term sustainability, all while ensuring the continuity of essential services even in the face of future climate-related shocks. Furthermore, the introduction of EPM software enhances the efficiency of these systems by enabling detailed reporting on performance and budgets, managing project portfolios, and tracking progress, thereby directly contributing to mitigating climate impacts, enhancing WASH services, and ensuring long-term sustainability even in the face of ongoing and new disasters. The project addresses immediate needs and contributes to a sustainable future, reinforcing the importance of integrating environmental considerations into development strategies.

Yet, in the Yemen case study (See Annex 3), the IOM's use of innovative technologies for water management underscores a forward-looking approach to integrating humanitarian aid with long-term development and is critical to addressing the unique complications that come from severe water scarcity in conflict and fragile contexts. Adaptation, sustainability, and resilience require carefully balancing immediate needs with the long-term needs of both human health and ecological balance. These technological solutions provide critical data for sustainable water use, including exploring aquifers' natural and artificial recharge during short but intense rainfall periods and enhancing water supply systems through low-cost, appropriate technology. Yemen's focus on improving the construction of extraction and irrigation systems, coupled with the extension of the groundwater monitoring program to key wells, highlights the crucial role of technology in enhancing water security and resilience.

This comprehensive integration of advanced technologies not only aids in the immediate response to water scarcity but also contributes to the sustainability and efficiency of water management strategies. Such initiatives are pivotal in reducing the environmental impact of water use, illustrating the crucial link between sustainable resource management and broader humanitarian and development goals.

Some new innovative technologies promise to be critical developments in the long-term adaptation, sustainability, and resilience of communities across the Middle East. The BluElephant System case study (see Annex 5), developed by Jotem Water Solutions in the Netherlands, introduces a ground-breaking, climate-resilient water management solution with the potential to address critical, community-level challenges around environmental pollution and water scarcity challenges in the face of climate change. The development of a decentralized water treatment system demonstrates the role new technological solutions can play in meeting the urgent needs of communities, particularly in areas where centralized wastewater plants are either at capacity or non-existent. The BluElephant System provides low-energy, cost-effective water treatment by recycling wastewater for irrigation and sanitary use, thus supporting the health of communities and the environment and enhancing the ability of communities to adapt to climate change while ensuring sustainable WASH outcomes. The project's efficiency and low maintenance make it a scalable solution for wastewater management, emphasizing the importance of innovative approaches in achieving sustainable WASH outcomes, especially critical in regions facing the compounded challenges of climate change and resource scarcity. Moreover, by enabling smaller communities to manage and repurpose scarce water supplies safely without the need for large, centralized infrastructure, BluElephant underscores the transformative potential of technology in fostering community ownership, sustainability, and environmental neutrality, thereby aligning directly with the principles of sustainable development and peacebuilding within the HDP framework.

Resource Management and Institutional Considerations

If countries across the Middle East are to successfully mitigate the current pressures of climate change and its growing implications, significant attention must be paid to developing and strengthening water resource management policies and other WASH-related institutions. This recommendation emphasizes the importance of institutional development within the WASH sector to address climate change pressures. Although institutional weaknesses have allowed the amplification of climate change consequences across the Middle East region, efforts can now be meaningful and effective in not only mitigating current deficits but also focusing on local and national assets, meaningful progress toward sustainable water resources, and WASH service resilience can be built.

Grieve, Panzerbieter, and Rück (2023) outline in the JOF that the first key steps local and national institutions need to take to address locally assessed weaknesses in WASH services are to,

- Mainstream the nexus approach in WASH programs at the national and subnational level, especially in protracted and recurrent crisis contexts.
- Create an enabling environment in key countries for implementing the WASH nexus approach, leveraging existing planning architecture and coordination structures where possible.
- Adopt flexible and sustainable financing strategies into protracted and recurrent crisis contexts, and leverage financing to achieve the SDGs.
- Advocate for predictable, un-earmarked, and flexible multi-year funding and the development of innovative financing and forecasting models.
- Continue to build key evidence, learning, and capacity (Grieve 2023a, 60–61; Grieve, Panzerbieter, and Rück 2023, 33).

First, the focus must be placed on institutional service providers over areas of water resource management and WASH services that lack the critical capacity to absorb and take future shocks and be strengthened with resilience programing. Resilience programing of this nature increases institutional capacity to anticipate shock and reduce local reliance on national resources. These anticipatory structures allow for the incremental strengthening of local WASH systems to address disrupted services in shorter time frames, creating sustainability and resilience (Grieve 2023a; Grieve, Panzerbieter, and Rück 2023).

Building on the foundation of institutional leadership and collaborative capacity facilitated by the JOF, there's a pressing need for a deeper commitment to locally led resilience and sustainability programing in all WASH services. Successful implementation requires developing and implementing coordinated management and adaptation mechanisms that ensure local, national, and international accountability. Water resources and WASH services must be well connected to monitoring systems that allow for fast and accurate decision-making, especially leading up to or during anticipated and unanticipated shocks. This will likely require legislative changes across the Middle East region and governments to reprioritize and realign obligations and commitments to meet better the current and future pressures climate change and water scarcity will pose. Local, national, and international resources will be critical to ensure that best practices and new technologies are fully utilized and locally maintained, thereby maximizing local resilience capabilities and ensuring long-term sustainability (ACWUA 2023). These suggested legislative changes to empower local and marginalized groups in climate adaptation efforts. There is ultimately a need for greater local and national leadership across the Middle East to address institutional weaknesses in a way that builds resilience and the ability to act proactively to address future climate change shocks, highlighting the need for leadership to overcome institutional inadequacies and prepare for climate impacts.

In the Yemen case study, the IOM's collaboration with Groundwater Relief to manage groundwater resources directly strengthens local water authorities' capacities, setting a foundation for resilient institutional responses to water scarcity. Meanwhile, in the Iraq case study, UNICEF's integration of green technologies into schools and healthcare facilities upgrades infrastructure and fosters systemic change towards sustainability, embedding resilience within local governance. Lastly, Jotem Water Solutions' BluElephant case study enhances community-level institutional capacities by providing sustainable wastewater management technologies and ensuring local stakeholders are prepared for environmental management and water sanitation. Each intervention reflects a strategic approach to enhancing the capabilities of institutions and communities to navigate environmental and humanitarian challenges sustainably. Linking institutional strengthening to the HDP nexus and strategies for combating climate change, these initiatives underscore the critical role of enhancing WASH outcomes through improved governance, capacity building, and sustainable practices, ensuring communities across the Middle East can navigate and thrive in the face of environmental challenges.

Nexus Funding and Sustainability

Up until now, we have argued that adopting a nexus mentality anchored in longer-term solutions, cross-collaboration, and planning is essential to address the impacts of climate change in the Middle East. However, as was highlighted in our interview with Karine Deniel, attracting funding for these interventions is difficult, complicated, and in some instances, near impossible. Denial's argued that

> There needs to be a push from both humanitarian and development actors to build better rather than quicker. However, the Humanitarian Response Plan, for example, only creates one year of funding at a time, which raises questions about the need for multi-year funding to support longer-term and more sustainable interventions that meet at the intersection between humanitarian and development actors.
>
> (Deniel 2023)

Funding presents a significant, if not the most significant, challenge for ensuring a sustainable and resilient transition across the HDP nexus in the Middle East, especially when considering the complexity of climate change and water scarcity. Financial structures often define how and where humanitarian and development programs are designed and implemented and can impede successful transitions across the HDP nexus. Yet we argue that effectively adopting a nexus mentality can help reduce long-term program costs and increase funding for climate change adaptation, mitigation, and WASH services. Although we acknowledge that these long-term benefits can require a short-term increase in humanitarian capital and a reduction in humanitarian reaction time, long-term considerations are required. Tillett et al. (2020) argued that building resilient WASH services in fragile contexts "requires predictable, programmatic, longer-term funding that WASH actors can rely on, independently of unpredictable shifts in the broader context. Such reliable funding streams must incentivize long-term planning and programming for sustainable outcomes, rather than short-term 'beneficiary' outputs" (Tillett et al. 2020, 61).

Humanitarian funding tends to be short term and crisis driven, emphasizing immediate lifesaving interventions such as emergency WASH services. However, addressing the root causes of water and sanitation challenges in the Middle East often requires sustained, long-term investments in infrastructure, capacity building, and governance. Climate-related challenges exacerbate existing WASH vulnerabilities, resulting in heightened competition for financial support to address urgent needs while simultaneously building climate-resilient infrastructure and practices. Securing funding for the transition from emergency response to development becomes challenging due to shifting funding priorities from short-term relief to long-term solutions (ICVA 2022). Climate-resilient WASH infrastructure is crucial to mitigate the impacts of climate change on water availability and quality. Climate financing mechanisms, when available, often prioritize sectors like energy or agriculture, leaving a gap in funding for critical climate-resilient WASH initiatives in the Middle East (SIWI and UNICEF 2023). Transitioning from immediate humanitarian relief to longer-term developmental projects in the WASH sector is complex. It involves redirecting resources, adjusting strategies, and ensuring a smooth handover from humanitarian actors to development organizations. This transition requires careful planning, coordination, and, most importantly, secure funding mechanisms to bridge the gap between emergency response and sustainable development.

Climate change introduces unpredictability in weather patterns and disasters, such as erratic rainfall and extreme weather events. This unpredictability can strain budgets and resources as humanitarian organizations must allocate funds quickly for emergency responses, potentially diverting resources from long-term development efforts to address evolving climate-related challenges in the Middle East (IPCC 2012). Integrating climate resilience into WASH programing involves added complexities, including the need for vulnerability assessments, climate risk reduction measures, and adaptation strategies. These complexities translate into increased costs and resource requirements, making it challenging to secure funding for comprehensive climate adaptation measures within WASH projects (Adger et al. 2009).

Donors have specific priorities and reporting requirements for their funding, often influenced by their geopolitical interests or strategic objectives.

Aligning these priorities with the dual objectives of addressing immediate humanitarian needs and building climate-resilient WASH systems in the Middle East can be challenging as it may require donor coordination and flexibility in funding mechanisms (Blaikie et al. 2004). Coordinating donors across humanitarian and development sectors is challenging due to different funding cycles, objectives, and priorities. Achieving synergy between humanitarian WASH interventions and climate-resilient development projects requires enhanced coordination mechanisms that promote collaboration among various stakeholders, donors, and implementing agencies in the Middle East (Macrae and Harmer 2004). In adopting the nexus mindset, donors must move beyond rhetoric, enhance inclusivity by listening to the affected populations, and deliver flexible, multi-year funding to align with the dynamic nature of crises in the region (ICVA 2022; Land and Hauck 2022).

These and other challenges create structures that prevent local and national institutional decision-makers from fully utilizing local capacity and assets. With the need for more sustainable funding for the operation and maintenance of WASH systems, scheduled activities are often canceled despite capacity and plans in place. If there was more financial sustainability in institutional systems, the needed skills and technical support could easily be mobilized to address other challenges. Making this point, Anna Rubert highlighted that

> We have institutions in Iraq, Syria, and Lebanon that are aware of the challenges, and have plans yet are pushed to have an emergency approach to face rising challenges (notably during conflict), but they are not being carried out due to a lack of financial capacity. The region potentially has all the needed capacity to find solutions. However, budgets assigned to WASH are reduced and losing value and are used primarily for salaries.
>
> (Rubert 2023)

Yet these salaries must be increased and adequate, incentivizing local skilled labor to remain in critical local leadership positions and preventing brain drain and the loss of local capacity. Correcting these weaknesses is essential to long-term sustainability. It sheds light on the huge challenge of transitioning from humanitarian funding and support by the international community towards longer-term funding and government support. In our interview with Denis Vanhontegem, he argued that a deeper "look at how the state budgets are allocated to ensure the sustainability of the services" is necessary for long-term climate change response and specified that subsides, along with other funding structures, were inadequate for long-term response (Vanhontegem 2023).

Financial sustainability is a central and recurring concern in addressing the financial needs required to ensure a seamless transition from short- to long-term WASH solutions. The region's reliance on state subsidies for urban water supply systems further amplifies the financial challenges, often hindering the establishment of cost recovery mechanisms vital for sustainability and resilience. The frequent need for effective water tariffs, taxes, and fees and the reluctance of authorities to prioritize WASH funding reflect the political nature of financial decisions. This, in turn, affects the capacity to invest in infrastructure expansion, maintenance, and quality improvements that are more innovative and effective in combating climate change. While awareness campaigns and innovative financing models hold potential, financial constraints permeate every aspect of the region's WASH landscape. The result is a precarious situation where budgets assigned to WASH are dwindling, undermining the sector's capacity to provide effective services and amplifying the vulnerabilities faced by marginalized populations affected by climate change. The region's funding variability creates a paradox: financial issues are the greatest challenge to effective climate change adaptation, mitigation programing, and WASH services (Malerba 2021). Yet, as a report by the UNICEF recently stated, "increasing WASH resilience may require higher upfront investments before reaping medium and longer-term benefits" (UNICEF 2020, 7). These dueling realities exacerbate the problems, preventing the establishment of sustainable options that could address long-term WASH concerns effectively. Mitigation efforts are simpler and cheaper than adaptation and resilience building and, therefore, receive a greater share of time and resources, pushing the long-term consequences of water scarcity and climate change off on future populations who are often the most marginalized (Malerba 2021).

Financial challenges in the Middle East present a critical obstacle to addressing WASH needs, especially considering the intricate web of climate change, crises, and the transition across the HDP nexus. While the challenges are intertwined with other aspects of the region's WASH landscape, financial instability emerges as the linchpin undermining progress. The need for financial sustainability, allocation, and adaptation becomes paramount to ensure the resilience of WASH systems. Blended finance and public-private partnerships represent opportunities to bring critical new funding streams but will require significant changes in local and national policy and priorities (UNICEF and GWP 2022). Without substantial efforts to secure consistent funding, the region's ability to provide essential WASH services to its populations amidst the ever-evolving challenges remains uncertain.

Conclusion

For the foreseeable future, climate change and its impacts will be the greatest threat to stability and peace in the Middle East (Alaaldin 2022; Howard et al. 2016; Malerba 2021). As we have argued here, climate change is a threat multiplier. It increases the length and impact of conflicts, hinders and obscures pathways to peace, and destroys environmental protection capabilities (United Nations 2020). In the Middle East, as with other areas of the world, the greatest pressure comes from acute water scarcity. As droughts and flooding increase, it becomes harder to manage scarce resources in a way that mitigates the current impacts of climate change while also effectively building local, national, and regional adaptation capabilities with sufficient strength to absorb future shocks (Alaaldin 2022; Baxter et al. 2022; Malerba 2021; Norwegian Red Cross 2019; UNICEF 2020; Waha et al. 2017). These failures place substantial pressure on the poorest and most marginalized populations in society, increasing economic inequalities and decreasing the potential for self-reliance, self-sufficiency, and sustainable livelihoods (Waha et al. 2017). There is a direct correlation between climate-related droughts causing increases in water scarcity, agricultural failures, and economic hardships that ultimately lead to poverty, civil unrest, and outright conflict (Abel et al. 2019). Younger populations, marginalized groups, and those with the least adaptive capacity disproportionally bear the consequences of climate change. Lacking the resources and capacity to adapt locally, these groups often migrate to urban centers, which in and of itself can be a climate-related shock on urban water and resource infrastructure, which, far too often, lacks the capacity and resiliency to take the added pressure of sudden population increases (Malerba 2021; Norwegian Red Cross 2019).

Climate change is an increasingly complex, long-term problem requiring significant changes in how HDP actors think about and design interventions. The Middle East, in particular, faces the compounding effects of extreme water scarcity coupled with climate-related fragility, violence, and conflict. The protracted nature of conflict and crisis in the Middle East requires humanitarian and development actors and organizations to adjust traditional mandates to allow for earlier development participation in fragile contexts while ensuring earlier and more effective transitions across the humanitarian and development divide. The NRC (2023) report also suggests that the transitionary space between the humanitarian and development sectors, termed the "grey zone," is increasingly important in addressing and solving protracted crises like climate change. The report suggests that as disasters and crises grow in length and scope, traditional humanitarian and development mandates and coordination mechanisms may be inadequate at building effective long-term adaptation, sustainability, and resilience against future climate change shocks (NRC 2023).

Transitioning across the humanitarian and development divide will require a stronger reliance on dual-mandate organizations, which are uniquely set to provide lifesaving humanitarian action while building long-term sustainability and resilience for the future. In protracted contexts like those caused by climate change, the NRC argues that "a sole reliance on repeated, one-off, short-term emergency assistance" cannot produce long-term, effective solutions (NRC 2023, 87). Rather, the NRC states that protracted crises require

Complementary sustainable interventions that contribute to strengthening local capacities and ensuring a more dignified response for affected communities. Humanitarian, or double-hatted organisations, were also found to have the context knowledge, networks, access, and operational experience to serve as the preferred partners for interventions that bridge the gap between humanitarian and development assistance.

(NRC 2023, 87)

Adaptation, sustainability, and resilience require local buy-in and long-term funding in traditionally short-term contexts. When led by local decision-makers, adaptation to water scarcity and climate change will have the best chance of success if built on the positive empowerment of local assets. Critically, HDP actors must act more as facilitators than designers, creating the best opportunities for long-term sustainability and resilience in climate-specific action.

References

A4EP. 2021. "Future Course for a Grand Bargain 2.0: Time to Walk Side by Side." Alliance for Empowering Partnership. https://resourcecentre.savethechildren.net/node/19011/pdf/a4ep-future-of-gb-statement.pdf.

Abel, Guy J., Michael Brottrager, Jesus Crespo Cuaresma, and Raya Muttarak. 2019. "Climate, Conflict and Forced Migration." *Global Environmental Change* 54: 239–49. https://doi.org/10.1016/j.gloenvcha.2018.12.003.

ACWUA. 2023. "6th Annual Arab Water Week." https://www.sri-2030.org/blog-post/report-on-the-6th-arab-water-week.

Adger, W. Neil, Suraje Dessai, Marisa Goulden, Mike Hulme, Irene Lorenzoni, Donald R. Nelson, Lars Otto Naess, Johanna Wolf, and Anita Wreford. 2009. "Are There Social Limits to Adaptation to Climate Change?" *Climatic Change* 93 (3): 335–54. https://doi.org/10.1007/s10584-008-9520-z.

Agenda for Change. 2023. "WASH Agenda for Change – Clean Water & Sanitation Organization." Agenda for Change. October 19, 2023. https://washagendaforchange.org/.

Akacha, Khaled. 2023. "Empower a Woman with Water and She Can Change Her City: A Focus on Mena." Cities Alliance, hosted by UNOPS.

Akther, Shahida, and Golam Mahbubul Alam. 2020. "Climate Change Causing Political Instability in the Middle East Region: A Critical Analysis." *Journal of Society & Change* XIV, 2 (June) 75–91.

Al Jazeera English. 2023. "ICRC: A Lack of Humanitarian Funding in Middle East Is Hurting the Communities Most in Need." June 1, 2023. https://www.youtube.com/watch?v=QQAMXQAO65o.

Al-Basha, Farah. 2023. Interview: Conducted by Mariëlle Snel and Reed Power on February 15, 2023.

Al-Salehi, Maha. 2023. "AMWAJ | Yemen: Groundwater Depletion and Possible Solutions." April 18, 2023. https://amwaj-alliance.com/yemen-groundwater-depletion-and-possible-solutions/.

Alaaldin, Ranj. 2022. "Climate Change May Devastate the Middle East. Here's How Governments Should Tackle It." Brookings. March 14, 2022. https://www.brookings.edu/articles/climate-change-may-devastate-the-middle-east-heres-how-governments-should-tackle-it/.

Baghdadi, Ahmad. 2022. "MENA's Role to Lead the World in Climate Education." WIRED Middle East. December 13, 2022. https://wired.me/science/menas-role-to-lead-the-world-in-climate-education/.

Bakchan, Amal, Miriam E. Hacker, and Kasey M. Faust. 2021. "Resilient Water and Wastewater Infrastructure Systems through Integrated Humanitarian-Development Processes: The Case of Lebanon's Protracted Refugee Crisis." *Environmental Science & Technology* 55 (9): 6407–20. https://doi.org/10.1021/acs.est.0c05630.

Abu Baker, Mohammad A., Nigel Reeve, April A. T. Conkey, David W. Macdonald, and Nobuyuki Yamaguchi. 2017. "Hedgehogs on the Move: Testing the Effects of Land Use Change on Home Range Size and Movement Patterns of Free-Ranging Ethiopian Hedgehogs." *PloS One* 12 (7): e0180826. https://doi.org/10.1371/journal.pone.0180826.

Barakat, Sultan, and Sansom Milton. 2020. "Localisation across the Humanitarian-Development-Peace Nexus." *Journal of Peacebuilding & Development* 15 (2): 147–63. https://doi.org/10.1177/1542316620922805.

Barbiche, Jean Christophe. 2023. Interview: Conducted by Mariëlle Snel and Reed Power on March 31, 2023.

Bastable, Andy. 2023. Interview: Conducted by Mariëlle Snel and Reed Power on February 14, 2023.

Batchelor, Charles, Stef Smits, and A. J. James. 2011. *Adaptation of WASH Services Delivery to Climate Change and Other Sources of Risk and Uncertainty (Thematic Overview Paper 24)*. The Hague, the Netherlands: IRC International Water and Sanitation Centre. http://www.irc.nl/top24.

Baxter, Louisa, Catherine R. McGowan, Sandra Smiley, Liliana Palacios, Carol Devine, and Cristian Casademont. 2022. "The Relationship between Climate Change, Health, and the Humanitarian Response." *The Lancet* 400 (10363): 1561–63. https://doi.org/10.1016/S0140-6736(22)01991-2.

Bayram, Mustafa, and Çağlar Gökırmaklı. 2020. *The Future of Food*. Cambridge Scholars Publishing. Newcastle upon Tyne, England.

Blaikie, Piers, Terry Cannon, Ian Davis, and Ben Wisner. 2004. *At Risk: Natural Hazards, People's Vulnerability and Disasters*. 2nd ed. London: Routledge. https://doi.org/10.4324/9780203714775.

Blind, Peride. 2019. "Humanitarian SDGs: Interlinking the 2030 Agenda for Sustainable Development with the Agenda for Humanity." UN Department of Economic and Social Affairs (DESA) Working Papers 160. Vol. 160. UN Department of Economic and Social Affairs (DESA) Working Papers. https://doi.org/10.18356/a2d75e71-en.

Bluelephant. n.d. "Bluelephant. From Wastewater to Infinite Water." BluElephant – NL. Accessed September 7, 2023. https://bluelephant.global/.

Cantor, David James. 2023. "Divergent Dynamics: Disasters and Conflicts as 'Drivers' of Internal Displacement?" *Disasters* 48 (1): e12589. https://doi.org/10.1111/disa.12589.

Caravani, Matteo, Jeremy Lind, Rachel Sabates-Wheeler, and Ian Scoones. 2022. "Providing Social Assistance and Humanitarian Relief: The Case for Embracing Uncertainty." *Wiley* 40 (5). https://doi.org/10.1111/dpr.12613.

Center on International Cooperation. 2019. "The Triple Nexus in Practice: Toward a New Way of Working in Protracted and Repeated Crises." Center on International Cooperation.

Chambers, Robert. 1997. *Whose Reality Counts? Putting the First Last*. London: Intermediate Technology Publications.

———. 2006. "Transforming Power: From Zero-Sum to Win-Win?" *IDS Bulletin* 37 (6): 99–110.

———. 2017. *Can We Know Better? Reflections for Development.* Rugby: Practical Action Publishing. http://dx.doi.org/10.3362/9781780449449.

Chambers, Robert, and Gordon Conway. 1992. "Sustainable Rural Livelihoods: Practical Concepts for the 21st Century." IDS Discussion Paper 296, Brighton: IDS.

Choptiany, John Michael Humphries, Benjamin Ernst Graeub, Sinan Hatik, Daniele Conversa, and Samuel Thomas Ledermann. 2019. "Participatory Assessment and Adaptation for Resilience to Climate Change." *Consilience,* (21) (July): 17–31. https://doi.org/10.7916/consilience.v0i21.5723.

Cloern, James E., Paulo C. Abreu, Jacob Carstensen, Laurent Chauvaud, Ragnar Elmgren, Jacques Grall, Holly Greening, et al. 2016. "Human Activities and Climate Variability Drive Fast-Paced Change across the World's Estuarine–Coastal Ecosystems." *Global Change Biology* 22 (2): 513–29. https://doi.org/10.1111/gcb.13059.

Congressional Research Service. 2023. "Climate Change and Security in the Middle East and North Africa." In *Focus.* Congressional Research Service. https://crsreports.congress.gov/product/pdf/IF/IF11878.

Cooper, Brittany, Nikki L. Behnke, Ryan Cronk, Carmen Anthonj, Brandie Banner Shackelford, Raymond Tu, and Jamie Bartram. 2021. "Environmental Health Conditions in the Transitional Stage of Forcible Displacement: A Systematic Scoping Review." *The Science of the Total Environment* 762 (March): 143136. https://doi.org/10.1016/j.scitotenv.2020.143136.

Cop, Serdar, Uju Alola, and Andrew Alola. 2020. "Perceived Behavioral Control as a Mediator of Hotels' Green Training, Environmental Commitment, and Organizational Citizenship Behavior: A Sustainable Environmental Practice – Cop – 2020 – Business Strategy and the Environment – Wiley Online Library." July 2020. https://onlinelibrary.wiley.com/doi/10.1002/bse.2592.

Corwith, Anne, and Erin Sorensen. 2023. "Integrated WASH and Education." In *Addressing Conflict, COVID-19, and Climate Change: A Multisectoral Approach to Integrated WASH Programming,* edited by Mariëlle Snel and Nikolas Sorensen. Rugby: Practical Action Publishing Ltd.

Darbyshire, Eoghan, Leonie Nimmo, Doug Weir, Rana El-Hajj, and Sarah Gale. 2023. "Making Adaptation Work." International Committee of the Red Cross and Norwegian Red Cross. https://reliefweb.int/report/world/making-adaptation-work-addressing-compounding-impacts-climate-change-environmental-degradation-and-conflict-near-and-middle-east.

Deniel, Karine. 2023. Interview: Conducted by Mariëlle Snel and Reed Power on February 14, 2023.

Development Initiatives. 2023. *Global Humanitarian Assitance Report 2023.* https://devinit.org/resources/global-humanitarian-assistance-report-2023.

DRC. 2021. "DRC Strategy 2025." Danish Refugee Council. https://www.pro.drc.ngo/media/xghmuvdu/drc-strategy-2025-en-nov-2021.pdf.

Duenwald, Christoph, Yasser Abdih, Kerstin Gerling, Vahram Stepanyan, Abdullah Al-Hassan, Gareth Anderson, Anja Baum, et al. 2022. "Feeling the Heat: Adapting to Climate Change in the Middle East and Central Asia." *International Monetary Fund* 2022 (008). https://doi.org/10.5089/9781513591094.087.

Dutch Water Sector. 2022. "Launch of BluElephants in Palestinian Territories Creates Infinite Water Flow." September 7, 2022. https://www.dutchwatersector.com/news/launch-of-bluelephants-in-palestinian-territories-creates-infinite-water-flow.

El Hattab, Omar. 2023. Interview: Conducted by Mariëlle Snel and Reed Power on February 15, 2023.

El-Geressi, Yasmine. 2020. "Climate Change, Water Woes, and Conflict Concerns in the Middle East: A Toxic Mix." Earth Day. September 8, 2020. https://www.earthday.org/climate-change-water-woes-and-conflict-concerns-in-the-middle-east-a-toxic-mix/.

Ensor, Jonathan. 2011. *Uncertain Futures*. Rugby: Practical Action Publishing. http://www.jstor.org/stable/j.ctt1hj595r.

Fantini, Emanuele. 2019. "An Introduction to the Human Right to Water: Law, Politics, and Beyond." *WIREs Water* 7 (2): e1405. https://doi.org/10.1002/wat2.1405.

FAO. 2020. "Project Proposal – Decentralised Cooperative Water Governance for Food Security and Peace in Yemen." FAO.

Fonseca, Catarina, and Lesley Pories. 2017. "Financing WASH: How to Increase Funds for the Sector While Reducing Inequities: Position Paper for the Sanitation and Water for All Finance Ministers Meeting." In The Hague: IRC, water.org, Ministry of Foreign Affairs and Simavi. https://www.ircwash.org/resources/financing-wash-how-increase-funds-sector-while-reducing-inequalities-position-paper.

Gengler, Justin. 2016. "The Political Economy of Sectarianism in the Gulf." Carnegie Endowment for International Peace. August 29, 2016. https://carnegieendowment.org/2016/08/29/political-economy-of-sectarianism-in-gulf-pub-64410.

Ghali, Maghie. 2022. "How Water Cooperation Can Provide Geopolitical Stability in MENA." Al Jazeera. April 2, 2022. https://www.aljazeera.com/news/2022/4/2/blue-peace-how-water-access-can-provide-mena-stability.

Giovanis, Eleftherios, and Oznur Ozdamar. 2021. "The Transboundary Effects of Climate Change and Global Adaptation: The Case of the Euphrates-Tigris Water Basin in Turkey and Iraq." *SSRN Electronic Journal & İzmir Bakırçay University*. https://ssrn.com/abstract=4320746.

GJU. 2020. "Humanitarian Water, Sanitation and Hygiene (WaSH)." Text. German Jordanian University. December 8, 2020. https://www.gju.edu.jo/content/humanitarian-water-sanitation-and-hygiene-wash-12063.

Global WASH Cluster. 2019. "Delivering Humanitarian WASH as Scale, Anywhere and Any Time: Road Map for 2020–2025." WASH Cluster.

Gray, Ian, Harriette Purchas, Monica Ramos, Julie Bara, Ross Tomlinson, and Aliocha Salagnac. 2022. *Global WASH Cluster, Strategic Plan 2022–2025*. Geneva: United Nations Children's Fund (UNICEF). https://www.washcluster.net/sites/gwc.com/files/inline-files/Global_WASH_Cluster_Strategic%20Plan_2022_2025_FINAL_lowres.pdf.

Grieve, Timothy. 2023a. "A WASH Framework to Address Conflict, COVID-19, and Climate Change: Leveraging the Humanitarian–Development–Peace Nexus." In *Addressing Conflict, COVID-19, and Climate Change: A Multisectoral Approach to Integrated WASH Programming*, edited by Mariëlle Snel and Nikolas Sorensen. Rugby: Practical Action Publishing Ltd. 33–65.

———. 2023b. Interview: Conducted by Mariëlle Snel and Reed Power on February 14, 2023.

Grieve, Timothy, Thilo Panzerbieter, and Johannes Rück. 2023. "WASH Resilience, Conflict Sensitivity and Peacebuilding: Joint Operational Framework." Triple Nexus in WASH Initiative. https://www.washnet.de/en/triple-nexus-wash/joint-operational-framework/.

Harvey, Ben, Murry Burt, Franklin Golay, and Ryan Schweitzer. 2019. *UNHCR WASH Manual: Practical Guidance for Refugee Settings*. Geneva: UNHCR. https://cms.emergency.unhcr.org/documents/11982/31573/UNHCR+WASH+Manual/57769fdb-ce92-4410-b86d-1faa0b79d9a3.

Heidebroek, Denis. 2023. Interview: Conducted by Mariëlle Snel and Reed Power on May 3, 2023.

Howard, Guy, Roger Calow, Alan Macdonald, and Jamie Bartram. 2016. "Climate Change and Water and Sanitation: Likely Impacts and Emerging Trends for Action." *Annual Review of Environment and Resources* 41 (1): 253–76. https://doi.org/10.1146/annurev-environ-110615-085856.

IASC. 2021. "The Grand Bargain (Official Website) | IASC." March 10, 2021. https://interagencystandingcommittee.org/grand-bargain.

Ibrahim, Esmaeil. 2023. Interview: Conducted by Mariëlle Snel and Reed Power on May 2, 2023.

ICRC. 2023. "Facing the Impact of Climate Change and Armed Conflict in the Near and Middle East." Report. Middle East/Iraq; Middle East/Syria; Middle East/Yemen. https://www.icrc.org/en/document/report-impact-climate-change-and-armed-conflict-near-and-middle-east.

ICRC, and NRC. 2023. "Making Adaptation Work – Addressing the Compounding Impacts of Climate Change, Environmental Degradation and Conflict in the Near and Middle East." International Committee of the Red Cross and Norwegian Red Cross.

ICVA. 2017. "The Grand Bargain Explained: An ICVA Briefing Paper." ICVA. https://reliefweb.int/sites/reliefweb.int/files/resources/ICVA_Grand_Bargain_Explained.pdf.

——— 2022. "Advancing Nexus in the MENA Region Breaking the Silos: Research and Documentation of the State of Humanitarian Development-Peace (HDP) Nexus in the MENA Region." ICVA. https://reliefweb.int/report/world/advancing-nexus-mena-region-breaking-silos-research-and-documentation-state-humanitarian-development-peace-hdp-nexus-mena-region-july-2022.

International Monetary Fund. 2016. "Economic Diversification in Oil-Exporting Arab Countries." *Policy Papers* 16 (28). https://doi.org/10.5089/9781498345699.007.

IOM. 2023. "New Water Distribution Network in Ma'rib Improves Access to Safe Water for 15,000 People." ReliefWeb. January 23, 2023. https://reliefweb.int/report/yemen/new-water-distribution-network-marib-improves-access-safe-water-15000-people-enar.

IOM DTM. 2023. "Iraq Master List Report 130, May–August 2023 – Iraq | ReliefWeb." October 24, 2023. https://reliefweb.int/report/iraq/iraq-master-list-report-130-may-august-2023.

IPCC. 2012. *Managing the Risks of Extreme Events and Disasters to Advance Climate Change Adaptation: A Special Report of Working Groups I and II of the Intergovernmental Panel on Climate Change.* Edited by C. B. Field, V. Barros, T. F. Stocker, D. Qin, D. J. Dokken, K. L. Ebi, M. D. Mastrandrea, et al. Cambridge and New York: Cambridge University Press.

———. 2018. "Summary for Policymakers. In: Global Warming of 1.5°C. An IPCC Special Report on the Impacts of Global Warming of 1.5°C above Pre-Industrial Levels and Related Global Greenhouse Gas Emission Pathways, in the Context of Strengthening the Global Response to the Threat of Climate Change, Sustainable Development, and Efforts to Eradicate Poverty." Cambridge and New York: Cambridge University Press, 3–24. https://doi.org/10.1017/9781009157940.001.

Jagerskog, Anders. 2023. Interview: Conducted by Mariëlle Snel and Reed Power on April 5, 2023.

Jotem. n.d. "BluElephant." Jotem. Accessed September 7, 2023. https://www.jotem.nl/en/solutions/bluelephant.

Jurczak, P., L. Whitley, and G. Burrows. 2023. "Hydrogeological Study, Ma'rib, Yemen." Study. Prepared by IOM, Ground Water Relief.

Kamal, Islam, Magdi Fekri, Islam Abou El-Magd, and Nashwa Soliman. 2021. "Mapping the Impacts of Projected Sea-Level Rise on Cultural Heritage Sites in Egypt: Case Study (Alexandria)." 5.

Khashman, Khaldon. 2023. Interview: Conducted by Mariëlle Snel and Reed Power on May 9, 2023.

Kurtzer, Jacob. 2019. "Never More Necessary: Overcoming HumanitarianAccess Challenges." September. https://www.csis.org/analysis/never-more-necessary-over coming-humanitarian-access-challenges.

Land, Tony, and Volker Hauck. 2022. "HDP Nexus: Challenges and Opportunities for Its Implementation – Final Report November 2022." *European Commission*, November.

Lelieveld, J., P. Hadjinicolaou, E. Kostopoulou, J. Chenoweth, M. El Maayar, C. Giannakopoulos, C. Hannides, et al. 2012. "Climate Change and Impacts in the Eastern Mediterranean and the Middle East." *Climatic Change* 114 (3): 667–87. https://doi.org/10.1007/s10584-012-0418-4.

Macrae, Joanna, and Adele Harmer. 2004. "Chapter 1 Beyond the Continuum: The Changing Role of Aid Policy in Protracted Crises." In *Beyond the Continuum: The Changing Role of Aid Policy in Protracted Crises*, edited by Adele Harmer and Joanna Macrae. HPG Report 18, ODI. London 1–13.

Malerba, Daniele. 2021. "Climate Change." In *Handbook on Social Protection Systems*, edited by Esther Schüring and Markus Loewe. Cheltenham: Edward Elgar Publishing. https://www.elgaronline.com/view/edcoll/9781839109102/9781839109102. 00041.xml.

Mason, Nathaniel, and Beatrice Mosello. 2016. "Making Humanitarian and Development WASH Work Better Together." Overseas Development Institute.

McCracken, Melissa. 2012. "The Impact of the Water Footprint of Qat on Yemen's Water Resources." Tufts University. https://www.researchgate.net/publication/320720558_ The_Impact_of_the_Water_Footprint_of_Qat_on_Yemen's_Water_Resources.

Mena, Rodrigo, and Dorothea Hilhorst. 2022. "The Transition from Development and Disaster Risk Reduction to Humanitarian Relief: The Case of Yemen during High-intensity Conflict." *Disasters* 46 (4): 1049–74. https://doi.org/10.1111/disa.12521.

Mena, Rodrigo, Summer Brown, Laura E. R. Peters, Ilan Kelman, and Hyeonggeun Ji. 2022. "Connecting Disasters and Climate Change to the Humanitarian-Development-Peace Nexus." *Journal of Peacebuilding & Development* 17 (3): 324–40. https://doi.org/10.1177/15423166221129633.

Mosel, Irina, and Simon Levine. 2014. *Remaking the Case for Linking Relief, Rehabilitation and Development: How LRRD Can become a Practically Useful Concept for Assistance in Difficult Places*. London: Overseas Development Institute. https://www.odi.org/sites/odi.org.uk/files/odi-assets/publications-opinion-files/8882.pdf.

Nakamitsu, Izumi, Ahunna Eziakonwa-Onochie, John Ging, and Marta Ruedas. 2017. "Humanitarian-Development Nexus: What Is the New Way of Working?" Webinar, April 26. http://www.deliveraidbetter.org/webinars/humanitarian-development-nexus/.

Neira, Marco, Kamil Erguler, Hesam Ahmady-Birgani, Nisreen DaifAllah AL-Hmoud, Robin Fears, Charalambos Gogos, Nina Hobbhahn, et al. 2023. "Climate Change and Human Health in the Eastern Mediterranean and Middle East: Literature Review,

Research Priorities and Policy Suggestions." *Environmental Research* 216 (January): 114537. https://doi.org/10.1016/j.envres.2022.114537.

Norwegian Red Cross. 2019. *Overlapping Vulnerabilities: The Impacts of Climate Change on Humanitarian Needs*. Oslo: Norwegian Red Cross.

NRC. 2022. "NRC Global Strategy 2022–2025." Norwegian Refugee Council. https://www.nrc.no/globalassets/pdf/policy-documents/global-strategy-2022-2025/nrc_global-strategy-2022-2025_english.pdf.

———. 2023. "The Nexus in Practice: The Long Journey to Impact." Norwegian Refugee Council.

OCHA. 2017. "The New Way of Working." OCHA. https://www.unocha.org/sites/unocha/files/NWOW%20Booklet%20low%20res.002_0.pdf.

———. 2023a. "Yemen Humanitarian Response Plan 2023." United Nations Office for the Coordination of Humanitarian Affairs. https://reliefweb.int/report/yemen/yemen-humanitarian-response-plan-2023-january-2023-enar.

———. 2023b. "Global Humanitarian Overview 2024." December 8, 2023. https://humanitarianaction.info/document/global-humanitarian-overview-2024/article/economic-hardship-persists-increasingly-becoming-primary-driver-need.

OECD. 2017. "Humanitarian Development Coherence: World Humanitarian Summit – Putting Policy into Practice – The Commitments into Action Series." Guidelines. Organisation for Economic Co-operation and Development. https://www.oecd.org/development/humanitarian-donors/docs/COHERENCE-OECD-Guideline.pdf.

———. 2018. "Enhancing the Legal Framework for Sustainable Investment Lessons from Jordan." OECD Global Relations Middle East and North Africa. https://www.oecd.org/mena/competitiveness/Enhancing-the-Legal-Framework-for-Sustainable-Investment-Lessons-from-Jordan.pdf.

———. 2021. "Middle East and North Africa Investment Policy Perspectives." OECD. https://doi.org/10.1787/6d84ee94-en.

———. 2022a. *The Humanitarian-Development-Peace Nexus Interim Progress Review*. Paris: OECD Publishing. https://doi.org/10.1787/2f620ca5-en.

———. 2022b. "Youth at the Centre of Government Action." https://www.oecd-ilibrary.org/content/publication/bcc2dd08-en.

Penney, Veronica, and John Muyskens. 2023. "Here's Where Water Is Running out in the World – and Why." Washington Post. August 16, 2023. https://www.washingtonpost.com/climate-environment/interactive/2023/water-scarcity-map-solutions/.

Peskett, Matt. 2023. "Middle East Continues to Attract Multiple Vertical Farming Companies." *Vertical Farming Today* (blog). February 17, 2023. https://www.verticalfarmingtoday.com/features/middle-east-continues-to-attract-multiple-vertical-farming-companies.html.

Rama, Martina. 2017. "Linking Relief, Rehabilitation and Development (LRRD): Examples and Lessons Learned for the WASH Sector." *40th WEDC International Conference*, Loughborough. https://www.pseau.org/outils/ouvrages/hydroconseil_wedc_linking_relief_rehabilitation_and_development_lrrd_examples_and_lessons_learned_for_the_wash_sector_2017.pdf.

RBAS. 2013. "Water Governance in the Arab Region Managing Scarcity and Securing the Future." United Nations Development Programme, Regional Bureau for Arab States (RBAS). https://www.undp.org/arab-states/publications/water-governance-arab-region.

REACH. 2022. "Briefing Note: Humanitarian Impact of Water Shortages in Northeast Syria (April 2022) – Syrian Arab Republic | ReliefWeb." April 20, 2022. https://

reliefweb.int/report/syrian-arab-republic/briefing-note-humanitarian-impact-water-shortages-northeast-syria-april.

Rizkallah, Amanda, Justin Gengler, Kathleen Reedy, and Ami Carpenter. 2019. *Countering Sectarianism in the Middle East.* Edited by Jeffrey Martini, Dalia Kaye, and Becca Wasser. Santa Monica, CA: RAND Corporation. https://doi.org/10.7249/RR2799.

Rubert, Anna. 2023. Interview: Conducted by Mariëlle Snel and Reed Power on May 2, 2023.

Saadeh, Dalia, Issam A. Al-Khatib, and Stamatia Kontogianni. 2019. "Public–Private Partnership in Solid Waste Management Sector in the West Bank of Palestine." *Environmental Monitoring and Assessment* 191 (4): 243. https://doi.org/10.1007/s10661-019-7395-2.

Sanitation and Water for All. 2020a. "The SWA Framework." Sanitation and Water for All (SWA). January 30, 2020. https://www.sanitationandwaterforall.org/about/our-work/priority-areas.

———. 2020b. "Building Blocks." Sanitation and Water for All (SWA). January 31, 2020. https://www.sanitationandwaterforall.org/about/our-work/priority-areas/building-blocks.

Satterthwaite, David, Diane Archer, Sarah Colenbrander, David Dodman, Jorgelina Hardoy, Diana Mitlin, and Sheela Patel. 2020. "Building Resilience to Climate Change in Informal Settlements." *One Earth* 2 (2): 143–56. https://doi.org/10.1016/j.oneear.2020.02.002.

Save the Children. 2022. "Save the Children Global Strategy 2022–24." Save the Children. https://www.savethechildren.net/sites/www.savethechildren.net/files/2022-24%20Save%20the%20Children%20Global%20Strategy.pdf.

Schyns, Joep F., Arwa Hamaideh, Arjen Y. Hoekstra, Mesfin M. Mekonnen, and Marlou Schyns. 2015. "Mitigating the Risk of Extreme Water Scarcity and Dependency: The Case of Jordan." *Water* 7 (10): 5705–30. https://doi.org/10.3390/w7105705.

Sen, Amartya. 1985. "A Sociological Approach to the Measurement of Poverty: A Reply to Professor Peter Townsend." *Oxford Economic Papers* 37 (4): 669–76.

———. 2000. *Development as Freedom.* New York: First Anchor Books. https://search.library.wisc.edu/catalog/999977297202121.

———. 2001. "Economic Development and Capability Expansion in Historical Perspective." *Pacific Economic Review* 6 (2): 179–91. https://doi.org/10.1111/1468-0106.00126.

Sitati, A., E. Joe, C. Grayson, C. Jaime, E. Gilmore, and E. Galappaththi. 2021. "Climate Change Adaptation in Conflict-Affected Countries: A Systematic Assessment of Evidence | Discover Sustainability." SpringerLink. 2021. https://link.springer.com/article/10.1007/s43621-021-00052-9.

SIWI, and UNICEF. 2023. *Water Scarcity and Climate Change Enabling Environment Analysis for WASH: Middle East and North Africa.* Stockholm and New York: Stockholm International Water Institute (SIWI) and United Nations Children's Fund (UNICEF). https://siwi.org/publications/water-scarcity-and-climate-change-enabling-environment-for-wash/.

Snel, Mariëlle, and Nikolas Sorensen. 2021. *Bridging the WASH Humanitarian-Development Divide: Building a Sustainable Reality.* Rugby: Practical Action Publishing Ltd. https://practicalactionpublishing.com/book/2577/bridging-the-wash-humanitariandevelopment-divide.

———., eds. 2023a. *Addressing Conflict, COVID-19, and Climate Change: A Multisectoral Approach to Integrated WASH Programming.* Rugby: Practical Action Publishing Ltd.

———. 2023b. "Focusing on the Future: Integration of WASH around Conflict, COVID-19, and Climate Change." In *Addressing Conflict, COVID-19, and Climate Change: A Multisectoral Approach to Integrated WASH Programming*, edited by Mariëlle Snel and Nikolas Sorensen. Rugby: Practical Action Publishing Ltd. 175–185

Sorensen, Nikolas, and Mariëlle Snel. 2022. "The New Reality: Perspectives on Future Integrated WASH." *Waterlines* 41 (1): 65–80. https://doi.org/10.3362/1756-3488. 20-00007OA.

Srivastava, Shilpi, Jeremy Allouche, Roz Price, and Tina Nelis. 2022. "Bringing WASH into the Water–Energy–Food Nexus in Humanitarian Settings." Institute of Development Studies. February 17, 2022. https://www.ids.ac.uk/publications/bringing-wash-into-the-water-energy-food-nexus-in-humanitarian-settings/.

Stavi, Ilan, Joana Roque de Pinho, Anastasia K. Paschalidou, Susana B. Adamo, Kathleen Galvin, Alex de Sherbinin, Trevor Even, Clare Heaviside, and Kees van der Geest. 2022. "Food Security among Dryland Pastoralists and Agropastoralists: The Climate, Land-Use Change, and Population Dynamics Nexus." *The Anthropocene Review* 9 (3): 299–323. https://doi.org/10.1177/20530196211007512.

Tanner, Thomas, David Lewis, David Wrathall, Robin Bronen, Nick Cradock-Henry, Saleemul Huq, Chris Lawless, et al. 2015. "Livelihood Resilience in the Face of Climate Change." *Nature Climate Change* 5 (1): 23–26. https://doi.org/10.1038/nclimate2431.

The Economist. 2023. "Climate Change | News and Analysis from The Economist." The Economist. 2023. https://www.economist.com/climate-change.

Tillett, W., J. Trevor, J. Schillinger, and D. DeArmey. 2020. "Applying WASH Systems Approaches in Fragile Contexts: A Discussion Paper." WASH Cluster. https://www.rural-water-supply.net/en/resources/959.

Turnbull, Marilise, Charlotte Sterrett, and Amy Hilleboe. 2013. *Toward Resilience: A Guide to Disaster Risk Reduction and Climate Change Adaptation*. Rugby: Practical Action Publishing Ltd.

UN DESA. 2017. "SDG 6, Water and Sanitation | Department of Economic and Social Affairs." September 2017. https://sdgs.un.org/topics/water-and-sanitation.

UN ESCWA. 2019. "Moving towards Water Security in the Arab Region." United Nations Economic and Social Commission for Western Asia. January 2019. http://www.unescwa.org/publications/moving-towards-water-security-arab-region.

UNHCR. 2023. "Global Trends Report 2022." 2023. https://www.unhcr.org/global-trends-report-2022.

UNICEF. 2019a. *Water Under Fire: For Every Child, Water and Sanitation in Complex Emergencies*. New York: United Nations Children's Fund (UNICEF).

———. 2019b. *Water Under Fire Volume 1: Emergencies, Development and Peace in Fragile and Conflict-Affected Contexts*. New York: United Nations Children's Fund (UNICEF).

———. 2020. "UNICEF Guidance Note: How UNICEF Regional and Country Offices Can Shift to Climate Resilient WASH Programming." United Nations Children's Fund (UNICEF).

———. 2021. *The Climate Crisis Is a Child Rights Crisis: Introducing the Children's Climate Risk Index*. New York: United Nations Children's Fund (UNICEF). https://www.unicef.org/reports/climate-crisis-child-rights-crisis.

———. 2022. *Global WASH Cluster, Strategic Plan 2022–2025*. Geneva: United Nations Children's Fund (UNICEF). https://www.washcluster.net/sites/gwc.com/

files/inline-files/Global_WASH_Cluster_Strategic%20Plan_2022_2025_FINAL_lowres.pdf.

UNICEF and GWP. 2022. "Strategic Framework for WASH Climate Resilient Development." United Nations Children's Fund (UNICEF) and Global Water Partnership (GWP).

UNICEF and WHO. 2023. "Progress on Household Drinking Water, Sanitation and Hygiene 2000–2022: Special Focus on Gender." United Nations Children's Fund (UNICEF) and the World Health Organization (WHO). https://data.unicef.org/resources/jmp-report-2023/.

UNICEF Jordan and Economist Impact. 2022. "Tapped out: The Costs of Water Stress in Jordan." https://jordan.un.org/en/191091-tapped-out-costs-water-stress-jordan

United Nations. 1989. "Convention on the Rights of the Child." OHCHR. November 20, 1989. https://www.ohchr.org/en/instruments-mechanisms/instruments/convention-rights-child.

———. 2017. "The New Way of Working | Joint Steering Committee to Advance Humanitarian and Development Collaboration." United Nations. 2017. https://www.un.org/jsc/content/new-way-working.

———. 2018. *World Humanitarian Data and Trends 2018*. United Nations. https://www.un-ilibrary.org/content/books/9789210475990.

———. 2020. "Climate Change Exacerbates Existing Conflict Risks, Likely to Create New Ones, Assistant Secretary-General Warns Security Council." United Nations. July 24, 2020. https://press.un.org/en/2020/sc14260.doc.htm.

———. 2023. "With Climate Crisis Generating Growing Threats to Global Peace, Security Council Must Ramp up Efforts, Lessen Risk of Conflicts, Speakers Stress in Open Debate." June 13, 2023. https://press.un.org/en/2023/sc15318.doc.htm.

———. n.d. "THE 17 GOALS | Sustainable Development." Accessed November 20, 2023. https://sdgs.un.org/goals.

United States Agency for International Development. 2022. "U.S. Government Global Water Strategy 2022–2027." ReliefWeb. October 12, 2022. https://reliefweb.int/report/world/us-government-global-water-strategy-2022-2027.

UNRISD. 2020. "The Humanitarian-Development-Peace Nexus: Towards Differentiated Configurations." ReliefWeb. September 8, 2020. https://reliefweb.int/report/world/humanitarian-development-peace-nexus-towards-differentiated-configurations.

Vanhontegem, Denis. 2023. Interview: Conducted by Mariëlle Snel and Reed Power on February 15, 2023.

Waha, Katharina, Linda Krummenauer, Sophie Adams, Valentin Aich, Florent Baarsch, Dim Coumou, Marianela Fader, et al. 2017. "Climate Change Impacts in the Middle East and Northern Africa (MENA) Region and Their Implications for Vulnerable Population Groups." *Regional Environmental Change* 17 (6): 1623–38. https://doi.org/10.1007/s10113-017-1144-2.

WASH Alliance International. 2021. "Welcome to the FIETS Sustainability Portal." January 29, 2021. https://wash-alliance.org/our-approach/sustainability/.

Waslander, Jacob. 2023. Interview: Conducted by Reed Power on March 30, 2023.

Wehrey, Frederic, Justin Dargin, Zainab Mehdi, and Marwan Muasher. 2023. "Climate Change and Vulnerability in the Middle East." Carnegie Endowment for International Peace. July 6, 2023. https://carnegieendowment.org/2023/07/06/climate-change-and-vulnerability-in-middle-east-pub-90089.

Werners, Saskia E., Edward Sparkes, Edmond Totin, Nick Abel, Suruchi Bhadwal, James R. A. Butler, Sabine Douxchamps, et al. 2021. "Advancing Climate Resilient Development Pathways since the IPCC's Fifth Assessment Report." *Environmental Science & Policy* 126 (December): 168–76. https://doi.org/10.1016/j. envsci.2021.09.017.

WHO and UNICEF. 2022. "Joint Monitoring and Analysis for WASH, IPC, Environment, and Waste Management Services of the Schools in Iraq." 2022. https:// app.powerbi.com/view?r=eyJrIjoiYTdiN2JhZDEtNWRlOC00NDVlLTliMzgt YTJhOTZmYWM5ZTJiIiwidCI6ImY2MTBjMGI3LWJkMjQtNGIzOS04MT BiLTNkYzI4MGFmYjU5MCIsImMiOjh9.

Winterford, Keren, Deborah Rhodes, and Christopher Dureau. 2023. *A Strengths-Based Approach for International Development: Reframing Aid.* Rugby: Practical Action Publishing. https://doi.org/10.1080/09614524.2023.2247582.

Wong, Catherine, Stephen Gold, Samuel Rizk, and Cassie Flynn. 2020. "Re-Envisioning Climate Action to Sustain Peace and Human Security." UNDP. November 17, 2020. https://www.undp.org/blog/re-envisioning-climate-action-sustain-peace-and-human-security.

World Bank. 2016. *High and Dry: Climate Change, Water and the Economy.* Washington, DC: World Bank.

World Bank, ICRC, and UNICEF. 2021. "Joining Forces to Combat Protracted Crises: Humanitarian and Development Support for Water Supply and Sanitation Providers in the Middle East and North Africa." World Bank. https://www.icrc.org/en/ document/joining-forces-secure-water-and-sanitation-protracted-crises.

World Bank, World. 2018. "Water Scarce Cities: Thriving in a Finite World." April. https://doi.org/10.1596/29623.

World Humanitarian Summit. 2016. "Commitments to Action: The World Humanitarian Summit, Istanbul May 23–24, 2016." Agenda for Humanity. https:// agendaforhumanity.org/sites/default/files/resources/2017/Jul/WHS_commitment_ to_Action_8September2016.pdf.

World Vision. 2020. "Macro-Catchment Construction: Alleviating the Threat of Both Flash Flooding and Water Shortages in Western Afghanistan." World Vision Afghanistan. https://www.wvi.org/publications/brochure/afghanistan/macro-catchment-construction.

World Waternet. 2022. "Global Launch of BluElephants in Palestinian Territories Creates Infinite Water Flow." September 7, 2022. https://www.wereldwaternet.nl/en/ latest-news/2022/september/global-launch-of-bluelephants-in-palestinian-territories-creates-infinite-water-flow/.

Annex 1
Glossary

Climate Change Changes in weather patterns over an extended period of time, often measured over decades.

Climate Change Adaptation Refers to the proactive development and application of strategies, policies, and practices to minimize risks and seize opportunities related to climate change. The WASH sector focuses on modifying practices and infrastructure, like diversifying water sources and enhancing system resilience, to ensure sustained access to services despite changing climate conditions.

Climate Change Mitigation Approaches to WASH that are designed to be sustainable, addressing immediate needs and long-term environmental sustainability. In addition, actions are taken to reduce or prevent the emission of greenhouse gases, aiming to limit the magnitude or rate of long-term climate change.

Climate Resilience This concept highlights the capacity of systems and communities to effectively anticipate, absorb, and recover from climate variability and extremes, ensuring the continuity of essential services and safeguarding public health and ecosystem services.

Do No Harm A principle in humanitarian and development work aimed at avoiding unintended negative consequences of interventions on the affected populations and local contexts.

Handover The process of transferring responsibility and ownership of projects or programs from one group (often humanitarian actors) to another (typically local authorities or communities), ensuring sustainability and local ownership of development or humanitarian interventions.

Humanitarian, Development, and Peace Nexus (HDP Nexus) or Triple Nexus The approach that seeks to integrate humanitarian aids, development efforts, and peacebuilding initiatives to address immediate needs, promote sustainable development, and contribute to long-lasting peace and stability in conflict-affected regions (UNICEF 2022).

Institutional Arrangements The structured organization and coordination of entities, roles, and responsibilities within the WASH sector are crucial for effective service delivery and management of water resources.

Joint Operational Framework (JOF) A tool designed to assist in integrating resilience, conflict sensitivity, and peacebuilding into WASH programs, promoting a unified approach across humanitarian, development, and peace pillars.

Localization The practice of emphasizing local leadership, resources, and expertise in humanitarian and development interventions to enhance local ownership, accountability, and sustainability, ensuring actions are aligned with local needs and contexts throughout all phases of humanitarian response and development.

Middle East The Middle East is a region primarily in western Asia and northeastern Africa, characterized by its strategic geopolitical significance, rich cultural and historical heritage, and vast natural resources. For the sake of this book, our definition includes Pakistan and Afghanistan.

Nexus Thinking/Nexus Mindset An approach that encourages considering the interconnectedness and interdependencies across different sectors and disciplines, such as humanitarian, development, and peace efforts, to address complex challenges in a more integrated and holistic manner.

Protracted Crises Situations where humanitarian crises extend over many years, often requiring sustained humanitarian response and resources.

Resilience The ability of individuals, communities, or systems to withstand, adapt to, and recover from the impacts of hazards and climate change without compromising their long-term viability (Turnbull, Sterrett, and Hilleboe 2013).

Resilience Building Resilience and resilience building is the process of strengthening institutions and helping individuals, families, and communities prepare for shocks in a way that allows them to either bounce back to a sense of normalcy or to adapt and pivot in a way to still provide for and meet their needs.

Social Cohesion The extent to which a society's members and leaders cooperate, share values, and work together towards common goals, enhancing mutual trust and a sense of belonging in communities.

Strengths-Based Approach A strategy that focuses on individuals', groups', or communities' inherent strengths and resources as the foundation for addressing challenges and achieving goals rather than on deficits or problems.

Sustainability The ability of WASH services and systems to maintain their long-term operations and benefits over the long term, ensuring that WASH needs of communities are met without compromising the ability of future generations to meet their own needs.

Transitional The phase of shifting interventions from emergency, short-term humanitarian aid to long-term, sustainable development initiatives, focusing on building resilience and sustainability.

Water, Hygiene, and Sanitation (WASH) From a humanitarian perspective, WASH refers to providing clean water, adequate sanitation, and proper hygiene practices to ensure the well-being and health of affected populations in crises.

Water, Energy, Food Nexus   The complex and interdependent relationship among water availability, energy production, and food provision, highlighting the imperative for coordinated and sustainable management and planning across these sectors to ensure resilience and resource efficiency.

Water Scarcity   A critical issue exacerbated by climate change, especially in arid regions, impacting access to water for drinking, agriculture, and sanitation and contributing to conflict and displacement.

Annex 2
List of Interviewee Respondents

Andy Bastable is the WASH Specialist and Co-Lead of the Global Cluster Faecal Sludge Management Technical Working Group at the Oxford Committee for Famine Relief (OXFAM).

Farah Al-Basha is the Global WASH Support Officer at the International Organization for Migration (IOM).

Jean Christophe Barbiche is the Regional WASH Adviser for the Middle East, Latin America, and Asia for the Norwegian Refugee Council (NRC).

Karine Deniel is the WASH Specialist with the IHE Delft Institute for Water Education.

Omar El Hattab is the Chief of Water, Environment, and Sanitation with the United Nations Children's Fund's (UNICEF) Pakistan Country Office.

Tim Grieve is the former Senior Advisor for Emergency Water, Sanitation and Hygiene with the United Nations Children's Fund (UNICEF) and the former Senior Policy Expert for Water Sanitation and Public Health with the German WASH Network.

Denis Heidebroek is the Regional WASH Shelter and Settlement Advisor with the European Commission's Civil Protection and Humanitarian Aid Operations department (ECHO).

Esmail Ibrahim is a WASH and Climate Change Consultant and former Chief of Water, Sanitation and Hygiene with the United Nations Children's Fund's (UNICEF) Jordan Office.

Anders Jägerskog is the Program Manager for the Cooperation in International Waters in Africa (CIWA) Trust Fund and Transboundary Waters Focal Point at the World Bank.

Khaldon Khashman is the Secretary General at the Arab Countries Water Utilities Association (ACWUA).

Anna Rubert is the Regional Advisory and Environmental Expert with the International Committee of the Red Cross (ICRC) in Amman, Jordan.

Denis Vanhontegem is the Climate Change and Resilience Advisor for the Middle East and Eastern Europe with Save the Children International.

Jacob Waslander is the Envoy to the Middle East and North Africa for Water, Energy, and Food with the Netherlands Ministry of Foreign Affairs based in Amman, Jordan, and the former Senior Associate for Climate Finance and Adaptation with the World Resources Institute.

Annex 3

Implementation of Sustainable Groundwater Management in the Humanitarian Response in Yemen

International Organization for Migration (IOM)

Farah Al-Basha

Yemen is one of the most water scarce countries in the world, with 15.3 million Yemenis lacking clean water and sanitation facilities (OCHA 2023a). Several underlying causes are exacerbating Yemen's water crisis, including the proliferation of Qat as a cash crop, which consumes more than 40% of Yemen's total renewable water resources and 32% of all groundwater withdrawals (McCracken 2012). According to FAO, groundwater withdrawal is twice the recharge amount. This means that in less than 20 years, most of the aquifers in Yemen will not be productive (FAO 2020). The water access situation in Yemen is dire and worsening due to the ongoing conflict, economic collapse, fuel shortages, and lack of maintenance of water infrastructure. Marib lies on the edge of the Ramlat al-Sab'atayn desert, which receives five days of rain annually. As such, stormwater runoff is critical to recharge aquifer systems. Due to conflict, the population of both Marib and the region around the Dam of Marib has grown from approximately 300,000 to over 1.1 million since 2014. Some 800,000 IDPs have moved to the region seeking refuge and stability (OCHA 2023b). The aquifer systems supporting Marib and the surrounding region were already under pressure before the influx of IDP migration. Since the influx, scarce water resources coupled with a significant increase in demand have led to overexploitation, resulting in a rapid decline in groundwater levels, a measurable increase in salinity, and overall deterioration of water quality across wells in the city.

Interventions

Water resource management in humanitarian contexts in semi-arid and arid regions requires integrated and adaptive approaches to cope with water availability, demand scarcity, and variability. As the WASH leads humanitarian agency in Marib, the IOM, with its regional water partners, seeks to improve water management strategies while sustainably responding to the humanitarian crisis in Marib. This requires developing a strategy to ensure medium- and long-term water management outcomes. The first stage of this process has been to analyze available groundwater resources within the region and the impact new abstraction pressures have on available resources.

Due to the limited availability of data regarding groundwater resources and management in Marib City, the IOM has commissioned Groundwater Relief (GWR) to conduct an initial field investigation. Through this first phase, GWR analyzed the available information on the hydrogeology of the region (literature, remote sensing data, and IOM databases). Also, the IOM and GWR conducted field studies on crucial water infrastructure, irrigation surveys, and water quality. They established a pilot-scale groundwater monitoring program to start the baseline groundwater monitoring across Marib. Through this investigation, 3D geological software was used to map aquifer systems; however, more data are needed to define aquifer systems properly. Nonetheless, the data clearly outline that groundwater mining is taking place with repercussions on the region's quality and level of water (Jurczak, Whitley, and Burrows 2023).

This study on water management in Marib provides the first basis for estimating future water availability while identifying potential deficits or surpluses. It also helps evaluate the feasibility and sustainability of various options to increase water supply or reduce water demand, such as improving the construction of extraction and irrigation systems, implementing groundwater monitoring systems, and improving water efficiency. This study also serves as a baseline for developing regional plans for coordinating the actions of authorities, water stakeholders, and humanitarian actors to achieve security and resiliency while reducing conflict's impact on water scarcity in Marib.

Water management studies are essential in a context of high humanitarian complexity, such as that found in Marib. Yet, conflict management and security pose a significant barrier to success. The ongoing conflict has disrupted the delivery of humanitarian aid, damaged water infrastructure, displaced millions of people, and increased the risk for water workers and users to violence and landmines. Mobilizing teams of experts requires intense logistical work and coordination with authorities and parties to the conflict. The IOM partnered with the Deputy Governor of Marib and other related water authorities to obtain approval for the study and use of equipment before the arrival of the GWR team. In consultation with local Marib authorities, the IOM security teams were used to ensure the safety of all teams and stakeholders. Due to security concerns, some equipment, such as groundwater loggers and study locations, were not authorized.

Water management in humanitarian emergencies is complex, with WASH actors quickly providing water to affected populations, often without sufficient analysis of water resource availability, affecting long-term decisions. In Marib, the IOM has shifted from emergency solutions like water trucking to sustainable projects, aiming to quickly adapt humanitarian responses to medium- and long-term needs, thus linking humanitarian efforts to peacebuilding and resilience. For example, in Marib, the IOM has responded by developing rapid, durable projects to supply drinking water that end the population's dependence on water trucking, including improving and rehabilitating groundwater pumping systems, storage infrastructure, distribution network construction, and community water system management activities. The recent construction of the Al Sowayda Water Supply Project includes a 6-kilometer transmission pipeline. This 15-kilometer water distribution network serves four displacement sites, a pumping station, and a 300 cubic meter water storage tank. This response focused on long-term solutions in humanitarian contexts, which extended the lifecycle and effectiveness of this humanitarian project (IOM 2023). Going forward, better data on regional water systems are crucial for managing scarce resources and ensuring sustainability in Marib. Despite challenges, obtaining detailed information on water resources, as demonstrated by the IOM's work, is challenging yet achievable.

Annex 4

Greening in Schools and Health Care Facilities in Iraq

UNICEF

Ali Al-Khateeb

Water scarcity and the increasing effects of climate change significantly impact Iraq's ability to provide safe drinking water and sanitation, especially to children and other marginalized groups. In schools and healthcare facilities, climate change-related droughts and extreme heat have led to closures and service disruption, negatively affecting the future livelihood potential of individuals, families, and communities. This is particularly true in areas where cooling systems are unavailable. Better water resource management is necessary to mitigate the increasing risk of water scarcity and the inadequate prevention and treatment of polluted water supplies.

In Iraq, inadequate electricity supply is one of the country's largest challenges, which impacts healthcare facilities and schools' ability to supply life-saving services and continue to provide education to children, particularly in under-resourced areas. The WHO and UNICEF conducted a comprehensive survey of 19,301 schools and 3,624 healthcare facilities in Iraq and determined that 48% of schools lack basic water services, 35% lack basic sanitation services, and 50% lack hand hygiene services, while 31% of health care facilities lack basic water services, 52% lack basic sanitation services, and 38% lack had hygiene services (UNICEF 2021; WHO and UNICEF 2022).

In an aim to address the impact of climate change on WASH services in schools and health care facilities in Iraq, UNICEF began a program to increase climate resilient and gender response facilities through the use of green, sustainable WASH services through the use of solar panels in schools and HCFs, decreasing WASH services reliance on diesel generators installing smart water meters to measure consumption and leakages, and calculating the sum of carbon reduction due to solar generation. A child-friendly online platform was also developed to generate evidence of water consumption, water leakages, and carbon reduction (UNICEF 2021).

New technologies, such as solar paneling, are becoming cheaper and more available alternatives to older centralized power grids. These options give healthcare facilities, water treatment plants, and schools the potential to become self-sufficient at the community level while reducing greenhouse gas emissions and increasing long-term sustainability, even in the face of ongoing and new disasters. In 2023, UNICEF started a task force with the Iraqi government to begin the process of greening schools and HCFs, with a particular emphasis on children's spaces, as they are vital agents of change in delivering and maintaining new technology. UNICEF provided technical support to the task force in designing and implementing solar systems consisting of 12 450W solar panels that powered the water pumps and lights at the school's entry and in the latrines. This setup was then applied to 199 schools and four healthcare facilities.

Furthermore, UNICEF created enterprise performance management software to increase the system's efficiency. This software helps create reports on performance and budgets, as well as manage project portfolios and track progress. UNICEF also supported school and HCF selection to ensure they had suitable water systems, met energy requirements and were ready for school administration. The 203 solar systems reduced annual carbon dioxide emissions by 1,240 tons. Finally, UNICEF helped 50 youth (nine female) participate in on-site training on the installation of the solar system and helped many of them subsequently obtain green jobs in the private sector as well as protection from sexual abuse.

Installation of green, solar systems at school not only provided opportunities to decentralize electrical and water systems but also provided valuable educational opportunities to teach students and youth about new technologies and renewable energy, gain hands-on experience in monitoring and maintaining units as well as teaching about water scarcity, water conservation and inspire action to conserve water resources. Additionally, more emphasis was placed on local community ownership of future water-related challenges.

Annex 5

BluElephant – Wastewater Treatment System in Palestine

Jotem Water Solutions

Bert Jansen

The BluElephant (Bluelephant n.d.; Jotem n.d.) is an infinite wastewater treatment system being developed by Jotem Water Solutions in the Netherlands and Palestine. Through various public-private partnerships, Jotem Water Solutions aims to use the BluElephant as a sustainable, environmentally safe, cost-effective, and community-driven solution to future water issues and aims to have the first units available in 2024. The system utilizes a Membrane Bioreactor (MBR) and a series of other filtration techniques to purify wastewater for irrigation and sanitary use. The MBR system utilizes anaerobic, aerobic, and anoxic zones to purify water naturally. It can also be equipped with active coal and ultraviolet water purifiers and can purify between 5m^3 and 6m^3 of water daily. Depending on water usage, each unit can support the daily water needs of between 80 and 160 people.

Central to the BluElephant's design are portability, simple maintenance, and low power consumption. Each unit is compactly designed (L × W × H: 2.2 × 2.2 × 2.5 meter) so that two units with their feed tanks and grind pumps can be mounted in one standard 20ft HC container and function as one system. Engineering has been designed so that all parts can be easily replaced from the outside of the unit, and the unit has a self-cleaning system that utilizes retail cleaning products. A low level of engineering is needed to be trained to maintain the system, requiring a one-day-long course to become fully capable of making future repairs. Finally, the system has been designed to run on a nominal 500W of power, briefly peaking while the pump runs at 2 kW. As such, one BluElephant can easily be plugged into a local power supply or run off a few solar panels and batteries, allowing each unit to be run independently and off the grid if necessary.

The water discharged from each unit meets all standards for sanitary use and as discharge to open water sources and irrigation systems and is over 97% efficient. As of February 2023, three units were being tested in the Netherlands, and three were tested in the West Bank in Palestine (Dutch Water Sector 2022; World Waternet 2022) near two hospital locations. The first units are anticipated to become available in 2024. As world populations increase, water becomes scarce and more contested, and surface water sources are becoming more polluted and harder to manage. Wastewater is a great concern as it can contaminate clean water sources if discharged or disposed of improperly.

BluElephant technology aims to be a solution for smaller communities to manage and repurpose, safely, already scarce water supplies in a way that doesn't require building and maintaining large, centralized waste-water plants. BluElephants can also be used as cheap, efficient solutions in areas where centralized wastewater plants have reached capacity. The decentralized nature and affordability of each BluElephant unit allow community owner-ship that is both sustainable and environmentally neutral.

Index

Printed in the United States
by Baker & Taylor Publisher Services